MONOGRAPHS ON STATISTICS AND APPLIED PROBABILITY

General Editors

D. R. Cox and D. V. Hinkley

Probability, Statistics and Time
M. S. Bartlett

The Statistical Analysis of Spatial Pattern
M. S. Bartlett

Stochastic Population Models in Ecology and Epidemiology
M. S. Bartlett

Risk Theory
R. E. Beard, T. Pentikäinen and E. Pesonen

Bandit Methods
D. A. Berry and B. Fristedt

Residuals and Influence in Regression
R. D. Cook and S. Weisberg

Point Processes
D. R. Cox and V. Isham

Analysis of Binary Data
D. R. Cox

The Statistical Analysis of Series of Events
D. R. Cox and P. A. W. Lewis

Analysis of Survival Data
D. R. Cox and D. Oakes

Queues
D. R. Cox and W. L. Smith

Stochastic Modelling and Control
M. H. A. Davis and R. Vinter

Stochastic Abundance Models
S. Engen

The Analysis of Contingency Tables
B. S. Everitt

Introduction to Latent Variable Models
B. S. Everitt

Finite Mixture Distributions
B. S. Everitt and D. J. Hand

Population Genetics
W. J. Ewens

Classification
A. D. Gordon

Monte Carlo Methods
J. M. Hammersley and D. C. Handscomb

Identification of Outliers
D. M. Hawkins

Generalized Linear Models
P. McCullagh and J. A. Nelder

Distribution-free Statistical Methods
J. S. Maritz

Multivariate Analysis in Behavioural Research
A. E. Maxwell

Applications of Queueing Theory
G. F. Newell

Some Basic Theory for Statistical Inference
E. J. G. Pitman

Statistical Inference
S. D. Silvey

Models in Regression and Related Topics
P. Sprent

Sequential Methods in Statistics
G. B. Wetherill

(Full details concerning this series are available from the Publishers)

Generalized Linear Models

P. McCULLAGH
Department of Statistics
University of Chicago

J. A. NELDER, F.R.S.
Department of Mathematics
Imperial College of Science and Technology
London

LONDON NEW YORK
CHAPMAN AND HALL

First published 1983 by
Chapman and Hall Ltd
11 New Fetter Lane, London EC4P 4EE
Published in the USA by
Chapman and Hall
29 West 35th Street, New York NY 10001
Reprinted 1984, 1985

Printed in Great Britain by the
University Press, Cambridge

ISBN 0 412 23850 0

British Library Cataloguing in Publication Data

McCullagh, P.
Generalized linear models.—(Monographs on statistics and applied probability)
1. Linear models (Statistics)
I. Title II. Nelder, J. A. III. Series
519.5 H61.25

ISBN 0-412-23850-0

Library of Congress Cataloging in Publication Data

McCullagh, P. (Peter), 1952–
Generalized linear models.

(Monographs on statistics and applied probability)
Bibliography: p.
Includes index.
1. Linear models (Statistics) I. Nelder, John A.
II. Title. III. Series.
QA276.M38 1983 519.5 83-7171
ISBN 0-412-23850-0

Contents

Preface xii

1 Introduction 1
1.1 Background 1
1.1.1 The problem of looking at data 2
1.1.2 Theory as pattern 3
1.1.3 Model fitting 4
1.1.4 What is a good model? 5
1.2 The origins of generalized linear models 7
1.2.1 Notation 7
1.2.2 Classical linear models 7
1.2.3 R. A. Fisher and the design of experiments 8
1.2.4 Dilution assay 9
1.2.5 Probit analysis 10
1.2.6 Logit models for proportions 11
1.2.7 Log–linear models for counts 11
1.2.8 Inverse polynomials 12
1.2.9 Survival data 13
1.3 Scope of the rest of the book 13
1.4 Bibliographic notes 14

2 An outline of generalized linear models 15
2.1 Processes in model fitting 15
2.1.1 Model selection 15
2.1.2 Estimation 17
2.1.3 Prediction 18
2.2 The components of a generalized linear model 18
2.2.1 The generalization 19
2.2.2 Likelihood functions for generalized linear models 20

2.2.3 Link functions 23
2.2.4 Sufficient statistics and canonical links 24
2.3 Measuring the goodness of fit 24
2.3.1 The discrepancy of a fit 24
2.3.2 The analysis of deviance 26
2.4 Residuals 28
2.4.1 Pearson residual 28
2.4.2 Anscombe residual 29
2.4.3 Deviance residual 30
2.5 An algorithm for fitting generalized linear models 31
2.5.1 Justification of the fitting procedure 32

3 Models for continuous data with constant variance 35
3.1 Introduction 35
3.2 Error structure 36
3.3 Systematic component (linear predictor) 37
3.3.1 Continuous covariates 38
3.3.2 Qualitative covariates 38
3.3.3 Dummy variates 40
3.3.4 Mixed terms 40
3.4 Model formulae for linear predictors 41
3.4.1 The dot operator 42
3.4.2 The + operator 42
3.4.3 The crossing (*) and nesting (/) operators 43
3.4.4 Operators for removal of terms 44
3.4.5 Exponential operator 45
3.5 Aliasing 45
3.5.1 Intrinsic aliasing with factors 48
3.5.2 Aliasing in two-way tables 49
3.5.3 Extrinsic aliasing 51
3.5.4 Functional relations in covariates 52
3.6 Estimation 53
3.6.1 The maximum-likelihood equations 53
3.6.2 Geometrical interpretation 54
3.6.3 Information 55
3.6.4 Model fitting with two covariates 56
3.6.5 The information surface 59
3.6.6 Stability 59

3.7 Tables as data 60
3.7.1 Empty cells 61
3.7.2 Fused cells 62
3.8 Algorithms for least squares 63
3.8.1 Methods based on the information matrix 63
3.8.2 Direct decomposition methods 65
3.8.3 Extension to generalized linear models 68
3.9 Selection of covariates 68
3.10 Bibliographic notes 71

4 Binary data 72
4.1 Introduction 72
4.1.1 Binary responses 72
4.1.2 Grouped data 73
4.1.3 Properties of the binomial distribution 74
4.1.4 Models for binary responses 75
4.1.5 Retrospective studies 77
4.2 Likelihood functions for binary data 79
4.2.1 Log likelihood for binomial data 79
4.2.2 Quasi-likelihoods for over-dispersed data 80
4.2.3 The deviance function 81
4.3 Asymptotic theory 81
4.3.1 Parameter estimation 81
4.3.2 Asymptotic theory for grouped data 82
4.3.3 Asymptotic theory for ungrouped data 84
4.3.4 Residuals and goodness-of-fit tests 85
*4.4 Conditional likelihoods 87
4.4.1 Conditional likelihood for 2×2 tables 87
4.4.2 Parameter estimation 89
4.4.3 Combination of information from several 2×2 tables 91
4.4.4 Residuals 92
4.5 Examples 92
4.5.1 Habitat preferences of lizards 92
4.5.2 Effect of a drug on cardiac deaths 98
4.6 Bibliographic notes 100

5 Polytomous data 101
5.1 Introduction 101
5.1.1 Measurement scales 102

5.1.2 Models for ordinal scales of measurement 102
5.1.3 Models for nominal scales of measurement 105
5.2 Likelihood functions for polytomous data 106
5.2.1 Log likelihood for multinomial responses 106
5.2.2 Quasi-likelihood functions for over-dispersed data 107
5.2.3 Asymptotic theory 109
5.3 Examples 109
5.3.1 A tasting experiment 109
5.3.2 Pneumoconiosis among coalminers 112
*5.4 Conditional and other likelihoods for the elimination of nuisance parameters 116
5.4.1 Proportional-odds models 117
5.4.2 Proportional-hazards or continuation-ratio models 120
5.4.3 Log–linear models 121
5.4.4 Discussion 124
5.5 Bibliographic notes 124

6 Log–linear models 127
6.1 Introduction 127
6.2 Likelihood functions 128
6.2.1 The Poisson log-likelihood function 130
6.2.2 Over-dispersion 131
6.2.3 Asymptotic theory 133
6.3 Examples 133
6.3.1 A biological assay of tuberculins 133
6.3.2 A study of ship damage 136
6.4 Log–linear models and multinomial response models 140
6.4.1 Comparison of two or more Poisson means 141
6.4.2 Multinomial response models 142
6.4.3 Discussion 143
6.5 Multiple responses 144
6.5.1 Conditional independence 144
6.5.2 Models for dependence 145
6.6 Bibliographic notes 147

7 Models for data with constant coefficient of variation 149
7.1 Introduction 149
7.2 The gamma distribution 150
7.3 Generalized linear models with gamma-distributed observations 153
7.3.1 The variance function 153
7.3.2 The deviance 153
7.3.3 The canonical link 154
7.3.4 Multiplicative models: log link 156
7.3.5 Linear models: identity link 156
7.3.6 Estimation of the dispersion parameter 157
7.4 Examples 158
7.4.1 Car insurance claims 158
7.4.2 Clotting times of blood 162
7.4.3 Modelling rainfall data using a mixture of generalized linear models 164

8 Quasi-likelihood models 168
8.1 Introduction 168
8.2 Definition 168
8.3 Examples of quasi-likelihood functions 169
8.4 Properties of quasi-likelihood functions 171
8.5 Parameter estimation 172
8.6 Examples 173
8.6.1 Incidence of leaf blotch on barley 173
8.6.2 Estimation of probabilities based on marginal frequencies 175
8.6.3 Approximating the hypergeometric likelihood 177
8.7 Bibliographic notes 180

9 Models for survival data 181
9.1 Introduction 181
9.2 Survival and hazard functions 181
9.3 Proportional-hazards models 183
9.4 Estimation with censored survival data and a survival distribution 183
9.4.1 The exponential distribution 185
9.4.2 The Weibull distribution 185

9.4.3 The extreme-value distribution 186
9.4.4 Other distributions 186
9.5 Example 186
9.5.1 Remission of leukaemia 186
9.6 Cox's proportional-hazards model 189
9.6.1 The treatment of ties 189
9.6.2 Numerical methods 190
9.7 Bibliographic notes 192

10 Models with additional non-linear parameters 193
10.1 Introduction 193
10.2 Parameters in the variance function 193
10.3 Parameters in the link function 195
10.3.1 One link parameter 196
10.3.2 More than one link parameter 198
10.3.3 Transformations of data vs transformations of fitted values 198
10.4 Non-linear parameters in the covariates 199
10.5 Examples 201
10.5.1 The effects of fertilizers on coastal Bermuda grass 201
10.5.2 Assay of an insecticide with a synergist 204
10.5.3 Mixtures of drugs 206

11 Model checking 209
11.1 Introduction 209
11.2 Techniques in model checking 209
11.3 The raw materials of model checking 210
11.4 Tests of deviations in particular directions 211
11.4.1 Checking a covariate scale 211
11.4.2 Checking the link function 212
11.4.3 Checking the variance function 212
11.4.4 Score tests for extra parameters 214
11.5 Visual displays 216
11.5.1 Checking form of covariate 217
11.5.2 Checking form of link function 218
11.5.3 Checking the variance function 218
11.6 Detection of influential observations 218
11.7 Bibliographic notes 221

12 Further topics 222
12.1 Second-moment assumptions 222
12.2 Higher-order asymptotic approximations 222
12.3 Composite link functions 223
12.4 Missing values and the EM algorithm 224
12.5 Components of dispersion 225

Appendices 229
A The asymptotic distribution of the Poisson deviance for large μ 229
B The asymptotic distribution of the binomial deviance 231
C The asymptotic distribution of the deviance: second-order assumptions 233
D The asymptotic distribution of the deviance: a higher-order approximation 237
E Software 239

References 241

Author index 251

Subject index 254

Preface

This monograph deals with a class of statistical models that generalizes classical linear models to include many other models that have been found useful in statistical analysis. These other models include log–linear models for the analysis of data in the form of counts, probit and logit models for data in the form of proportions (ratios of counts), and models for continuous data with constant proportional standard error. In addition, important types of model for survival analysis are covered by the class.

An important aspect of the generalization is the presence in all the models of a *linear predictor* based on a linear combination of explanatory or stimulus variables. The variables may be continuous or categorical (or indeed mixtures of the two), and the existence of the linear predictor means that the concepts of classical regression and analysis-of-variance models, insofar as they refer to the estimation of parameters in a linear predictor, carry across directly to the wider class of model. In particular, the ideas underlying factorial models, including those of additivity, interaction, polynomial contrasts, aliasing, etc., all appear in the wider context.

Generalized linear models have a common algorithm for the estimation of parameters by maximum likelihood; this uses weighted least squares with an adjusted dependent variate, and does not require preliminary guesses to be made of the parameter values.

The book is aimed at applied statisticians and postgraduate students in statistics, but will also be useful, in part at least, to undergraduates and to numerate biologists. More mathematical sections are marked with asterisks and may be omitted at first reading. Some mathematics have been relegated to the first four appendices, while the fifth contains information on computer software for the fitting of generalized linear models. The book requires knowledge of matrix theory, including generalized inverses, together with basic ideas of probability theory, including orders of magnitude in probability. As far as possible, however, the development is self-contained, though necessarily fairly condensed, because of the constraints on a monograph in this series. Further reading is given in bibliographic sections of chapters, and the theory is illustrated with a diverse set of worked examples.

We are grateful to Professor Zidek of the University of British Columbia and to the Natural Sciences and Engineering Research Council, Canada, for the opportunity to undertake an intensive spell of writing. For permission to use previously unpublished data we wish to thank Dr Graeme Newell, Lloyds Register of Shipping, and Drs P. M. Morse, K. S. McKinlay and D. T. Spurr. We are grateful to Miss Lilian Robertson for her careful preparation of the manuscript.

London and Harpenden
1983

P. McCullagh
J. A. Nelder

CHAPTER 1

Introduction

1.1 Background

In this book we consider a class of statistical models that is a natural generalization of classical linear models. The class of *generalized linear models* includes as special cases linear regression and analysis-of-variance models, logit and probit models for quantal responses, log–linear models and multinomial response models for counts. It is shown that the above models share a number of properties such as linearity and have a common method for computing parameter estimates. These common properties enable us to study generalized linear models as a single class rather than as an unrelated collection of special topics.

Classical linear models and least squares began with the work of Gauss and Legendre (Stigler, 1981) who applied the method to astronomical data. Their data were usually measurements of continuous quantities such as the position and magnitude of the heavenly bodies and, at least in the astronomical investigations, the variability in the observations was largely the effect of measurement error. The Normal, or Gaussian, distribution was a mathematical construct developed to describe the properties of such errors; later in the nineteenth century the distribution was used to describe the variation of individuals in a biological population in respect of a character such as height, an application quite different in kind from its use for describing measurement error, and leading to the numerous biological applications of linear models.

Gauss introduced the Normal distribution of errors as a device for describing variability but he showed that the important properties of least-squares estimates depended not on Normality but on an assumption of constant variance. A closely related property applies to all generalized linear models. In other words, although we make reference at various points to standard distributions such as the

Normal, binomial, Poisson, exponential or gamma, the second-order properties of the parameter estimates do not depend on the assumed distributional form but only on the variance-to-mean relationship and uncorrelatedness. This is fortunate because one can rarely be confident that the assumed distributional form is necessarily correct.

Another strand in the history of statistics is the development of methods for dealing with discrete events rather than with continuously varying quantities. The enumeration of probabilities of configurations in cards and dice was a matter of keen interest in the eighteenth century, and from this grew methods for dealing with data in the form of counts of events. The basic distribution here is that named after Poisson, and it has been widely applied to the analysis of such data, which may refer to many diverse kinds of events: a famous example concerns unfortunate soldiers kicked to death by Prussian horses (Bortkewitsch, 1898); others might concern the infection of an agar plate in a pathology experiment, the presence or absence of disease in a medical patient, the arrival of a neutron from a source, and so on. Generalized linear models allow us to develop models for the analysis of counts analogous to classical linear models for continuous quantities. Closely related are the models for the analysis of data in the form of proportions, or ratios of counts, such as occur when, for example, disease incidence is related to factors such as age, social class, exposure to pollutants, and so on. Again generalized linear models supply the appropriate analogues to classical linear models for such data.

Some continuous measurements have non-Normal error distributions and the class of generalized linear models includes models useful for the analysis of such data. Before looking in more detail at the history of individual examples of generalized linear models, we make some general comments about statistical models and the part they play in the analysis of data, whether experimental or observational.

1.1.1 *The problem of looking at data*

Suppose we have a number of measurements (or counts) together with some associated structural information, such as the order in which they were collected, which measuring instrument was used, and other differing conditions under which the individual measurements were made. To interpret such data we search for a *pattern* in them, for example that one measuring instrument has produced consistently higher readings than another. Such a systematic effect is likely to be blurred by other variation of a more haphazard nature; such variation will be described in statistical terms and so summarized in terms of its mass behaviour.

Statistical models contain both elements, systematic and random effects, and their value lies in allowing us to replace the individual data values by a summary that describes their general characteristics in terms of a limited number of quantities. Such a reduction is certainly necessary, for the human mind, while it may be able to encompass say 10 numbers easily enough, finds 100 much more difficult, and will be quite defeated by 1000 unless some reducing process takes place.

Thus the problem of looking intelligently at numerical data demands the formulation of patterns which can in some way represent the data and allow their important characteristics to be described in terms of a limited number of quantities that the mind can encompass relatively easily.

1.1.2 *Theory as pattern*

We shall consider theories as generating patterns of numbers which in some sense can replace the data, and can themselves be described by a few numerical quantities. These quantities are termed *parameters*, and by giving them values specific patterns can be generated. Thus in the very simple model

$$y = \alpha + \beta x,$$

connecting two quantities y and x, the parameter pair (α, β) defines the straight-line relation connecting x and y. We now suppose some causal relation between x and y under which x is under experimental control and affects y which we can measure (ideally) without error. Then if we give x the values

$$x_1, \quad x_2, \quad \ldots, \quad x_n,$$

y will take values

$$\alpha + \beta x_1, \quad \alpha + \beta x_2, \quad \ldots, \quad \alpha + \beta x_n$$

for some values α and β. Clearly if we know α and β we can reconstruct the values of y exactly from those of x, so that given $x_1, x_2, \ldots, x_n$ the pair (α, β) is an exact summary of $y_1, y_2, \ldots, y_n$ and we can move between the data and parameters in either direction.

In practice, of course, we never measure the ys exactly, so that the relation between y and x is only approximately linear. We can still choose values of α and β, a and b say, which in some sense best describe the now approximate relation between y and x. The quantities $a+bx_1, a+bx_2, \ldots, a+bx_n$ which we denote by $\hat{y}_1, \hat{y}_2, \ldots, \hat{y}_n$ are now the *theoretical* or *fitted* values generated by the model and they do not exactly reproduce the original data $y_1, y_2, \ldots, y_n$. They represent a patterned set which approximates the data, and they can be summarized by the pair (a, b).

1.1.3 *Model fitting*

The fitting of the simple linear relation between the ys and the xs requires us to choose a pair (a, b) from all possible (α, β) that gives us the patterned set $\hat{y}_i$ closest to the data y_i. How can we define 'closest'? To do so requires some measure of discrepancy to be defined between the ys and the $\hat{y}$s, and this definition requires certain assumptions to be made about the variation in the ys that is not accounted for by the model. Classical least squares chooses $\Sigma_i(y_i - \hat{y}_i)^2$ as the measure of discrepancy. This formula has two implications. First, the simple summation of the individual terms $(y_i - \hat{y}_i)^2$, each depending on only one observation, implies that the measurements are all made on the same physical scale and suggests that the measurements are independent in some sense. Secondly, the use of a function of $y_i - \hat{y}_i$ implies that a given deviation carries the same weight irrespective of the value of $\hat{y}$. If we model the unaccounted-for variation in statistical terms, then the first property of independence becomes stochastic independence, and the second property is interpreted by requiring that the variance of the distribution of deviations be independent of the mean. Statisticians are accustomed to go further and to build into their model an explicit frequency distribution for the residual variation. In classical least squares this was the Normal (or Gaussian) distribution, in which

$$\text{frequency of } y \text{ given } \mu \propto \exp[-(y-\mu)^2/(2\sigma^2)], \qquad (1.1)$$

where σ is the standard deviation of the distribution and acts as a scale factor describing the 'width' of the errors when measured about the mean.

We can look at the function in (1.1) in two ways. If we regard it as a function of y for fixed μ (1.1) specifies the probability density of the observations. Alternatively, for a given observation y we may consider (1.1) as a function of μ giving the relative plausibility of different values of μ for the particular value of y observed. In the second form it becomes the likelihood function, whose importance was first stressed by R. A. Fisher. We notice that the quantity $-2l$, where l is the log likelihood, is, for a simple sample of n values, just

$$\frac{1}{\sigma^2}\sum_{i=1}^{n}(y_i-\mu_i)^2,$$

i.e. it is, apart from the factor σ^2, here assumed known, the same as the sum-of-squares criterion. As μ varies, $-2l$ takes its minimum value at $\mu = \bar{y}$, the arithmetic mean of the ys. For a more complex model (with μ varying from value to value) we define the closest set $\hat{y}$ to be that whose values maximize the likelihood (or equivalently minimize $-2l$). More generally, we can have regard not only to the point that minimizes $-2l$ but also to the shape of the likelihood surface in the neighbourhood of the minimum. This shape tells us, in Fisher's terminology, how much *information* there is about the parameter(s) of the model in the data, as well as the best-fitting value(s).

Reverting to our example of a linear relationship, we can plot on a graph with axes α and β the contours of equal discrepancy ($-2l$) for a given data set y. Here the contours are ellipses, similar and similarly situated with the maximum-likelihood estimates (a, b) at the centre. The information in the data on (α, β) is given by the curvature (Hessian) matrix; if the axes of the ellipses are not aligned with the (α, β) axes, this means that the estimates are correlated. Note that information is greatest in the direction where the curvature is greatest (see Fig. 3.7). The form of the information surface can often be estimated *before* an experiment is carried out, and this can be used to compute the experimental resources required to estimate parameters with a required accuracy. It will also show in which directions information is least, and this can be very valuable to the experimenter. Alas, such calculations are not made nearly often enough!

1.1.4 *What is a good model?*

To summarize, we aim in model fitting to replace our data $\mathbf{y}$ with a set of fitted values $\hat{\mathbf{y}}$ derived from a model, these fitted values minimizing some criterion such as the discrepancy measure $\Sigma_i(y_i-\hat{y}_i)^2$.

At first sight it might seem as though a good model is one that fits the data very well, i.e. that makes $\hat{\mathbf{y}}$ very close to $\mathbf{y}$. However, by including enough parameters in our model we can make the fit as close as we please, and indeed by having as many parameters as observations we can make the fit perfect. In so doing, however, we have achieved no reduction in complexity – produced no simple theoretical pattern to substitute for the ragged data. Thus *simplicity*, represented by *parsimony* of parameters, is also a desirable feature of a model; we do not include parameters that we do not need. Not only does a parsimonious model enable the analyst to think about his data, but one that is substantially correct gives better *predictions* than one that includes unnecessary extra parameters.

An important property of a model is its *scope*, i.e. the range of conditions over which it gives good predictions. Scope is hard to formalize, but easy to recognize, and intuitively it is clear that scope and parsimony are to some extent related. If a model is made to fit very closely to a particular set of data, it will not be able to encompass the inevitable changes that will be found to be necessary when another set of data relating to the same phenomenon is collected. Both scope and parsimony are related to *parameter invariance*, that is to parameter values that either do not change as some external condition changes or change in a predictable way.

Modelling in science remains, partly at least, an art. Some principles do exist, however, to guide the modeller. The first is that *all models are wrong*; some, though, are better than others and we can search for the better ones. At the same time we must recognize that eternal truth is not within our grasp. The second principle (which applies also to artists!) is not to fall in love with one model, to the exclusion of alternatives. Data will often point with almost equal emphasis at several possible models and it is important that the analyst accepts this. A third principle involves checking thoroughly the fit of a model to the data, for example by using residuals and other quantities derived from the fit to look for outlying observations, and so on. Such procedures are not yet fully formalized (and perhaps never will be), so that imagination is required of the analyst here as well as in the original choice of models to fit. A recent paper by Box (1980) attempts some formalization of the dual processes of model fitting and model criticism.

1.2 The origins of generalized linear models

1.2.1 *Notation*

This section deals with the origin of generalized linear models, describing various special cases which are now included in the class in approximately their historical order. First we need some notation: Data will be represented by a *data matrix*, a two-dimensional array in which the rows are experimental or survey *units* (e.g. plots in an agricultural field trial, patients in a medical survey, quadrats in an ecological study) and the columns are *variates* (e.g. measurements, characters, treatments, grouping factors, etc.). Some of the variates will be *dependent* variates, whose values are believed to be affected by other *covariates* (sometimes, and unfortunately, called independent variates). Tukey (1962) uses *response* and *stimulus* variates for the two classes. Covariates may be *quantitative* or *qualitative*, i.e. taking values, numeric or non-numeric, from a finite set. We shall refer to qualitative covariates as *factors*, and they may include grouping factors, or treatment factors in an experiment. Dependent variates may be continuous, or discrete (in the form of counts), or take the form of factors, taking one of a set of alternative values (see Chapter 5).

1.2.2 *Classical linear models*

In modern notation we have a column vector of observations $\mathbf{y} = \{y_1, y_2, \ldots, y_n\}^T$ and an $n \times p$ matrix $\mathbf{X}$ with values of p covariates $\mathbf{x}_1, \mathbf{x}_2, \ldots, \mathbf{x}_p$ as its columns. Unknown parameters $\boldsymbol{\beta} = \{\beta_1, \beta_2, \ldots, \beta_p\}$ are associated with the covariates, and given arbitrary values of the $\boldsymbol{\beta}$s we can define a vector of *residuals*

$$\mathbf{e} = \mathbf{y} - \mathbf{X}\boldsymbol{\beta}.$$

Legendre in 1805 first proposed estimating the βs by minimizing $\mathbf{e}^T\mathbf{e} = \Sigma_i e_i^2$ for variation in the βs. [Note that both Legendre and Gauss defined the residuals with opposite sign to that in current use, i.e. by $\mathbf{e} = \mathbf{X}\boldsymbol{\beta} - \mathbf{y}$.] Gauss in 1809, in an astronomical text, introduced the Normal distribution with zero mean and constant variance for the errors. Later, in his *Theoria Combinatiores* in 1821, he abandoned the Normal distribution in favour of a weaker assumption of constant variance only, showing that the estimates of β obtained from the least-squares criterion have minimum variance in the class of unbiased estimates. The generalization of this weaker assumption to generalized linear models as a whole was given by Wedderburn (1974), using the concept of quasi-likelihood (Chapter 8).

Classical linear models are rooted in astronomy, as has been mentioned above, and the associated method of least squares for the estimation of parameters goes back to Gauss. Most of the data analysed using this technique were of the *observational* kind, i.e. they originated from *observing* a system, such as the Solar System, without being able to *perturb* it, i.e. to experiment with it. The development of the theory of experimental design gave a new stimulus to linear models and is very much associated with R. A. Fisher and his co-workers.

1.2.3 *R. A. Fisher and the design of experiments*

Fisher began work at Rothamsted in 1919 and within 10 years he had, among other things, laid the foundations of the design of experiments, an area that was greatly developed by his successor, Yates, and others at Rothamsted. In particular Fisher stressed the value of *factorial experiments* in which many experimental factors were applied simultaneously instead of one at a time. Thus with two factors, each at two levels, the one-at-a-time design (a) in Fig. 1.1 was replaced by the factorial design (b).

Factorial design increased information per experimental unit, and the analysis led to *factorial models* in which the yields were considered

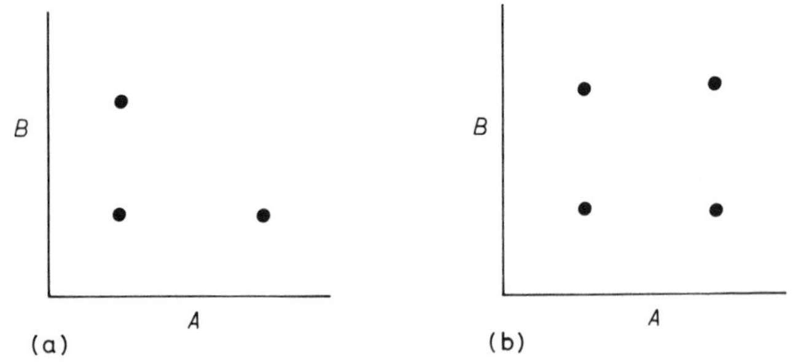

Fig. 1.1. (*a*) *Design for two factors, changing levels one at a time;* (*b*) *factorial design.*

as expressible as the sum of effects due to factors acting one at a time (main effects), two at a time (two-factor interactions), and so on. These factorial models are an important component of all generalized linear models; they are often called analysis-of-variance models, and contrasted with regression models for observational data with continuous covariates. We shall not make this distinction, and in fact will seek to unify the ideas embodied in these two extremes by including also terms in which the slopes defined by regression parameters are allowed to change with the level of some indexing factor.

Fisher's influence on the development of generalized linear models extends well beyond the models for factorial experiments and includes special models for the analysis of certain kinds of counts and proportions. We now consider some of the non-classical cases of generalized linear models which arose in the period 1922–1960.

1.2.4 *Dilution assay*

The original paper here is Fisher (1922, section 12.3). A solution containing an infective organism is progressively diluted, and at each dilution agar plates are 'streaked' and the number producing growth on the medium is counted. From the number of sterile plates at each dilution an estimate of concentration is made. The argument is as follows, where for simplicity we assume dilutions in powers of 2. The expected frequency of organisms at any stage can be written

$$\mu = N/2^x,$$

where x is the number of dilutions and N, which we wish to estimate, is the number of organisms in the original sample. Under random sampling the chance of a plate receiving $0, 1, 2, 3, \ldots$ organisms follows the Poisson distribution, so that the chance p of a plate remaining sterile is $e^{-\mu}$, the first term in the Poisson series. Thus $-\ln p = \mu$, where ln denotes the natural logarithm and so

$$\ln(-\ln p) = \ln N - x \ln 2.$$

If at dilution x we have r sterile plates out of n, then our measurement y is the fraction sterile ($= r/n$) and

$$\mathrm{E}(Y) = p.$$

However this time it is not p, the theoretical mean, which is linear in x, but a function

$$\eta = \ln(-\ln p),$$

for we can write

$$\eta = \alpha + \beta x,$$

where $\alpha = \ln N$ and $\beta = -\ln 2$.

Notice that here we have a slope β which is known *a priori*, an intercept α which is the logarithm of the quantity N which we want to estimate, and that it is not the theoretical mean p of the measurement y which is linear, but a function (here $\ln(-\ln p)$) of it. Fisher described the use of maximum likelihood to estimate p and also used his concept of information to show that another estimator, based solely on the total number of sterile plates over all dilutions, contained 87.7% of the information of the maximum-likelihood estimator. Nowadays we can use a computer to derive the maximum-likelihood estimator, and the model is just a special case of the class of generalized linear models. The estimation procedure is an early example of the use of maximum likelihood.

1.2.5 *Probit analysis*

This technique also arose in connection with bioassay, and the modern method of analysis dates from Bliss (1935). Several sets of organisms are given a set of doses of a stimulant (toxin, etc.) and the number, Y, responding is measured. The model relates the expected proportion p responding at dose x (measured on some suitable scale) by

$$p = \Phi(\alpha + \beta x),$$

when Φ is the Normal cumulative distribution function, and α and β are parameters to be estimated. Note that we cannot use $\Phi^{-1}(y/n)$ as the response variate in our analysis because it becomes infinite at $y = 0$ or n. Note also that again the postulated linearity is not between some function of y and x, but between a function of p, the *expected* proportion, and x. This is characteristic of generalized linear models, and reflects the fact that linearity is not a property of the data, but rather of the fitted (theoretical) values defined by the model. Note also that the variance of our observations (Y/n) is not constant as in classical least-squares models, but varies with the mean p, being zero at $p = 0$ or 1, and rising to a maximum at $p = 0.5$. Generalized linear models can allow variances of observations to be functions of the mean.

1.2.6 *Logit models for proportions*

Dyke and Patterson (1952) published an analysis of some cross-classified survey data of proportions responding in which they postulated a factorial model applying not to the expected proportions p themselves but to the logit or log odds $= \ln[p/(1-p)]$ of those proportions. They were successful in finding a simple model of this kind, though the fitting, which was done by hand, took several days whereas now it would take a second or two on a modern computer. Linear logistic models had earlier been used in the context of bioassay (see, for example, Berkson, 1944, 1951).

1.2.7 *Log–linear models for counts*

The analysis of counts has recently given rise to a large literature, mainly based on the idea of a log–linear model. In this model the two components of the classical least-squares model are replaced in the following way:

	Classical least squares	*Log–linear model*
Systematic effects	additive	multiplicative
Error distribution	Normal	Poisson

The Poisson distribution is the basic distribution for data in the form of counts, i.e. taking the discrete values $0, 1, 2, \ldots$. It has one parameter μ equal to the mean which must be non-negative, as compared to the parameter of the Normal distribution which is unrestricted. For this reason, an additive model for μ will usually be unsatisfactory, because the *linear predictor* $\eta = \Sigma\beta_i x_i$ may become negative, thus giving impossible values for η if we set $\mu = \eta$. The multiplicative model sets $\mu = e^{\eta}$ and this ensures that μ remains non-negative for all values of η. The ideas of a factorial model or a regression model remain as with the classical model, except that now the parameters β measure effects on the scale of log frequencies.

An extension of log–linear models occurs when one of the classifying factors is a response factor, i.e. we are measuring how the other factors change the relative frequencies of the levels of the response factor. Here the natural error distribution is a multinomial distribution; such a distribution, as is well known, can be regarded as a set of independent Poisson distributions, subject to the constraint that the total count is fixed (see Section 6.4). A suitable initial (or minimal) log–linear model can be fitted to the data to fix the fitted total counts at their given values; additional terms to test the effect of other factors on the response factor will then be fitted conditional on those totals being kept fixed (Nelder, 1974).

There can be no doubt that the introduction of log–linear models, another special case of the class of generalized linear models, has had a major impact on the analysis of counts. Details are given in Chapters 4 to 6.

1.2.8 *Inverse polynomials*

Polynomials have been widely used for expressing the shape of response curves in biological experiments. While they are easily fitted to data, they have disadvantages, the main one being that if extrapolated, their values become unbounded, and frequently negative, though the measurements themselves are generally required to be positive. In applications, however, it is common to find that a response approaches a plateau (or asymptote) as the stimulus increases, and polynomials do not have asymptotes. Hyperbolic response curves of the form

$$x/y = \alpha + \beta x$$

have been used in many contexts (e.g. the Michaelis–Menten equations of enzyme kinetics), and the inverse polynomials which were introduced in Nelder (1966) generalize this kind of response to include inverse quadratic terms and joint response functions for several variables. It is common to find errors where the proportional rather than the absolute standard error is constant, and this combination of error structure with the inverse response surface characterizes another subset of generalized linear models.

1.2.9 *Survival data*

In the last 15 years great interest has developed in models for survival in the context of medical and surgical treatment and for the analysis of failure times of manufactured components. Usually the data contain censored items, i.e. we know that the item or individual survived up to a certain time but we do not know the outcome beyond that. It has recently been shown (Aitkin and Clayton, 1980; Whitehead, 1980) how the analysis of such censored survival data can be put into the framework of generalized linear models, with the result that programs developed to fit generalized linear models can easily be adapted to the analysis of certain kinds of survival data.

1.3 Scope of the rest of the book

In Chapter 2 we outline the component processes in model fitting and describe the components of a generalized linear model, the definition of goodness of fit of a model to a set of data, a general method of fitting generalized linear models and some asymptotic theory of the statistical properties of the parameter estimates. Chapter 3 deals with classical models for continuous data, with a linear model for the systematic effects and errors with constant variance. Many of the ideas introduced here carry over with little or no change to the class of generalized linear models. The following three chapters describe generalized linear models that are relevant to data in the form of counts or proportions (ratios of counts). The random variation is described in terms of the Poisson, binomial or multinomial distributions and the systematic effects are assumed additive on a suitable scale which predicts relative frequencies that are always positive, or proportions lying between 0 and 1. These models include log–linear models, logit models for proportions, and models which underlie probit analysis. Chapter 7 introduces generalized linear models for continuous data where the coefficient of variation, rather than the variance, is assumed constant; examples are drawn from rainfall data and amounts in car insurance claims.

A major extension to the applicability of generalized linear models was made by Wedderburn (1974) when he introduced the idea of quasi-likelihood. This weakened the assumption that we assume that we know exactly the distribution of the random component in the model, and replaced it by an assumption about how the variance changed with the mean. Models based on quasi-likelihood are introduced informally where appropriate in earlier chapters, while in Chapter 8 a more systematic account is given of this important idea, which has many applications.

Medical research is much concerned with the analysis of survival times of individuals with different histories and undergoing different treatments. There is a close connection between the models underlying the analysis of survival data and generalized linear models, and in Chapter 9 this connection is explored.

It frequently happens that a model would fall into the class of generalized linear models if the values of certain parameters were known *a priori*. In the absence of such knowledge the fitting of a model of this kind involves extensions to the method of Chapter 2, and these are described in Chapter 10.

It has become increasingly recognized in recent years that a model fitted to data may be demonstrably inadequate from the evidence of the data themselves. Techniques which allow the systematic exploration of a fit and various kinds of deviation from it to be detected are outlined in Chapter 11. Finally in Chapter 12 we discuss briefly some further relevant topics, including the EM algorithm and composite link functions.

1.4 Bibliographic notes

The historical development of linear models and least squares from Gauss and Legendre to Fisher has previously been sketched. The use of the scoring method which amounts to iterative weighted least squares in the context of probit analysis is due to Fisher (1935) in an appendix to a paper by Bliss (1935). Details are given by Finney (1971). The term 'generalized linear model' is due to Nelder and Wedderburn (1972) who extend the scoring method to maximum-likelihood estimation in exponential families. Linear exponential families have been studied by Dempster (1971), Berk (1972), Barndorff-Nielsen (1978) and Haberman (1977). Special cases of linear exponential family models have been studied by Cox (1970) and by Breslow (1976).

CHAPTER 2

An outline of generalized linear models

2.1 Processes in model fitting

In Chapter 1 we considered briefly some of the *reasons* for model fitting as an aid to interpreting data. Before describing the form of generalized linear models we look now at the *processes* of model fitting, following closely the ideas of Box and Jenkins (1976), which they applied to time series. We distinguish three processes: (i) model selection, (ii) estimation and (iii) prediction (Box and Jenkins use 'model identification' in place of our 'model selection', but we wish to avoid any implication that a correct model can ever be known with certainty). In distinguishing these processes we make no assumption that an analysis consists of the successive application of each just once. In practice there are backward steps, false starts that have to be retracted, and so on. We now look briefly at some of the ideas associated with each of the three processes.

2.1.1 *Model selection*

Models that we select to fit to data will usually be chosen from a particular class, and this class must be broadly relevant to the kind of data we have if the model-fitting process is to be useful. An important characteristic of generalized linear models is that they assume *independent* (or at least *uncorrelated*) *observations*, so that, for example, data showing the autoregressions of time series are excluded; this assumption of independence is characteristic of the linear models of classical regression analysis, and is carried over to the wider class of generalized linear models. A second assumption about the error structure is that there is a single error term in the model; we thus exclude, for instance, models for the analysis of designed experiments of the kind giving rise to more than one error term, the simplest example being the split-plot design with its two error terms associated with between-main-plot and within-main-plot comparisons.

The restrictiveness of these two error assumptions is in practice somewhat less than might appear at first sight. For example, with multi-way tables of counts non-independence may arise because a marginal total needs to be fixed in the analysis. The analysis conditional on this fixed margin can in fact proceed as if the errors were independent. Similarly a within-groups analysis can ignore correlations induced by the grouping factor in a nuisance classification whose effects have to be eliminated before the effects of other (important) factors can be estimated.

The choice of scale of measurement is an important aspect of model selection – should we, for example analyse Y or $\ln Y$? The question is what characterizes a 'good' scale. In classical regression analysis a good scale has to combine constant variance and approximate Normality of errors with additivity of systematic effects. Now there is of course no *a priori* necessity that a scale with such desirable properties should exist, and it is not difficult to construct cases where it will not. Take, for example, the analysis of counts where the error structure is well approximated by a Poisson distribution and the systematic effects of some classifying factors are multiplicative in effect. If Y has a Poisson distribution then the square-root scale will give us approximate constancy of variance, while $Y^{\frac{2}{3}}$ will do better for producing approximate Normality; finally $\ln Y$ will be better for additivity of systematic effects. Clearly no one scale will produce all the desired effects. The extension of classical regression models to generalized linear models largely removes the scaling problem; Normality and constant variance are no longer a requirement for the error component while additivity of systematic effects can be specified to hold on a transformed scale if necessary.

There remains a major problem in model selection, namely the choice of x-variables (or *covariates*, as we shall call them) to be included in the systematic part of the model. There is a large literature on this problem in classical regression: We are given a number of possible covariates $x_1, x_2, \ldots, x_p$ and it is required to find a subset of these which is in some sense optimum for constructing a set of fitted values

$$\hat{\boldsymbol{\mu}} = \sum b_j \mathbf{x}_j.$$

Implicit in the strategies proposed is that there is a balance to be struck between the improvement in the goodness of fit to the data obtained by adding an extra term to the model, and the extra (and usually undesirable) increase in complexity introduced by such an addition. Note that even if we can define exactly what we mean by an optimum model in a given context, it is most unlikely that the data will indicate a clear winner among possible models, and we must expect that clustered round the 'optimum' model will be a set of others almost as good and not statistically distinguishable. Selection of covariates is further discussed in Section 3.9.

2.1.2 *Estimation*

Given a decision to select a particular model we then have to estimate the unknown parameters and obtain some measure of the accuracy with which we have estimated them. Estimation proceeds by defining a measure of goodness of fit between the data and a corresponding set of fitted values generated by the model, and choosing the parameter estimates as those that minimize the chosen goodness-of-fit criterion. We shall be concerned primarily with estimates obtained by maximizing the likelihood (or more conveniently the log likelihood) of the parameters given the data. If $f(y;\theta)$ is the frequency function under a given model for observation y with parameter θ, then the log likelihood, expressed as a function of the mean-value parameter $\mu = \mathrm{E}(Y)$, is given by $l(\mu;y) = \ln f(y;\theta)$. For a set of independent observations the log likelihoods are additive functions of the individual observations.

There are advantages in using as the goodness-of-fit criterion not the log likelihood $l(\boldsymbol{\mu};\mathbf{y})$ but a linear function of it, $-2[l(\boldsymbol{\mu};\mathbf{y})-l(\mathbf{y};\mathbf{y})]$ where $l(\mathbf{y};\mathbf{y})$ is the maximum likelihood achievable for an exact fit in which fitted values equal the data. Because $l(\mathbf{y};\mathbf{y})$ does not depend on the parameters, maximizing $l(\boldsymbol{\mu};\mathbf{y})$ and minimizing $-2[l(\boldsymbol{\mu};\mathbf{y})-l(\mathbf{y};\mathbf{y})]$, which for generalized linear models we call the *deviance*, are equivalent. Thus for classical regression with known variance σ^2, we have for one observation y with mean μ

$$f(y;\mu) = \frac{1}{\sqrt{(2\pi\sigma^2)}}\exp\left(-\frac{(y-\mu)^2}{2\sigma^2}\right),$$

so that

$$l(\mu;y) = -\tfrac{1}{2}\ln(2\pi\sigma^2)-(y-\mu)^2/(2\sigma^2).$$

Setting $\mu = y$ gives

$$l(y; y) = -\tfrac{1}{2}\ln(2\pi\sigma^2),$$

and so

$$-2[l(\mu; y) - l(y; y)] = (y-\mu)^2/\sigma^2.$$

2.1.3 *Prediction*

Prediction is concerned with the formation of derived quantities which are functions of the estimates and the associated fitted values, together with measures of their uncertainty. Thus in a quantal-response assay we measure the proportion of individuals responding to a range of dose levels. We then fit a model expressing how this proportion varies with dose, and from the model we predict the dose giving, for example, a 50% response rate. (In this context the word calibration is often used.) Prediction is commonly thought of in the context of time series, where it is required to predict the future behaviour of a system from observations of its past behaviour. However, the above example shows that prediction is wider in scope than this, time entering only indirectly, in that we are predicting the dose at which a further sample of the population of individuals used for testing would give a 50% response. Clearly the accuracy of prediction depends upon the model selected, and in turn, one criterion for selecting a model could be to maximize the accuracy of a particular prediction. Where experiments are concerned, further control is possible and covariate values can be chosen to optimize prediction. For an account of prediction as a unifying idea connecting the analysis of covariance and various kinds of standardization, see Lane and Nelder (1982).

2.2 The components of a generalized linear model

The class of generalized linear models is an extension of that of classical linear models, so that these latter form a suitable starting point. A vector of observations $\mathbf{y}$ of length N is assumed to be a realization of a vector of random variables $\mathbf{Y}$ independently distributed with means $\boldsymbol{\mu}$. The vector of means $\boldsymbol{\mu}$ constitutes the systematic part of the model and we assume the existence of covariates $\mathbf{x}_1, \mathbf{x}_2, \ldots, \mathbf{x}_p$ with known values such that

$$\boldsymbol{\mu} = \sum_{1}^{p} \beta_j \mathbf{x}_j, \tag{2.1}$$

where the βs are *parameters* whose values are usually unknown and have to be estimated from the data. If we let i index the observations then the systematic part of the model may be written

$$\mathrm{E}(Y_i) = \mu_i = \sum_1^p \beta_j x_{ij}; \qquad i = 1, \ldots, N, \tag{2.2}$$

where x_{ij} is the value of the jth covariate for observation i. In matrix notation (where $\boldsymbol{\mu}$ is $n \times 1$, $\mathbf{X}$ is $n \times p$ and $\boldsymbol{\beta}$ is $p \times 1$) we may write

$$\boldsymbol{\mu} = \mathbf{X}\boldsymbol{\beta},$$

where $\mathbf{X}$ is the *model matrix* and $\boldsymbol{\beta}$ is the vector of parameters. This completes the specification of the systematic part of the model.

The assumptions of independence and constant variance for the error term are strong and need checking, as far as is possible, from the data themselves. We shall consider techniques for doing this in Chapter 11. Similarly, the structure of the systematic part assumes that we know the covariates that influence the mean and can measure them effectively without error; this assumption also needs checking, as far as this is possible.

A further specialization of the model involves the stronger assumption that the errors follow a Gaussian or Normal distribution with constant variance σ^2.

We may thus summarize the classical linear model in the form:

Y are independent Normal variables with constant variance with

$$\mathrm{E}(\mathbf{Y}) = \boldsymbol{\mu} \qquad \text{where} \qquad \boldsymbol{\mu} = \mathbf{X}\boldsymbol{\beta}. \tag{2.3}$$

2.2.1 *The generalization*

To simplify the transition to generalized linear models, we shall rearrange (2.3) slightly to produce the following tripartite form:

1. The random component: $\mathbf{Y}$ has an independent Normal distribution with constant variance σ^2 and $\mathrm{E}(\mathbf{Y}) = \boldsymbol{\mu}$.
2. The systematic component: covariates $\mathbf{x}_1, \mathbf{x}_2, \ldots, \mathbf{x}_p$ produce a *linear predictor* $\boldsymbol{\eta}$ given by

$$\boldsymbol{\eta} = \sum_1^p \beta_j \mathbf{x}_j.$$

3. The *link* between the random and systematic components:

$$\boldsymbol{\mu} = \boldsymbol{\eta}.$$

This form introduces a new symbol η for the linear predictor and the third part then specifies that μ and η are in fact identical. If we write

$$\eta_i = g(\mu_i),$$

then $g(\ )$ will be called the *link function*. In this formulation, classical linear models thus have a Normal (or Gaussian) distribution in part 1 and the identity function for the link in part 3. Generalized linear models allow two extensions; first the distribution in part 1 may come from an exponential family and secondly the link function in part 3 may become any monotonic differentiable function. We look first at the extended distributional assumption.

2.2.2 *Likelihood functions for generalized linear models*

The form we use gives the scalar observation y a probability density function

$$f_Y(y;\theta,\phi) = \exp\{[y\theta - b(\theta)]/a(\phi) + c(y,\phi)\} \tag{2.4}$$

for some specific functions $a(\ )$, $b(\ )$ and $c(\ )$. If ϕ is known, this is an exponential family with canonical parameter θ. It may or may not be an exponential family if ϕ is unknown. Thus for the Normal distribution

$$\begin{aligned} f_Y(y;\theta,\phi) &= \frac{1}{\sqrt{(2\pi\sigma^2)}}\exp[-(y-\mu)^2/2\sigma^2] \\ &= \exp\{(y\mu - \mu^2/2)/\sigma^2 - \tfrac{1}{2}[y^2/\sigma^2 + \ln(2\pi\sigma^2)]\}, \end{aligned}$$

so that $\theta = \mu$, $b(\theta) = \mu^2/2$, $\phi = \sigma^2$, $a(\phi) = \phi$,

and
$$c(y,\phi) = -\tfrac{1}{2}[y^2/\sigma^2 + \ln(2\pi\sigma^2)].$$

Write $l(\theta,\phi;y) = \ln f_Y(y;\theta,\phi)$ for the log-likelihood function considered as a function of θ and ϕ, y being given. Then the mean and variance of Y can be derived easily from the well known relations

$$\mathrm{E}\left(\frac{\partial l}{\partial \theta}\right) = 0 \tag{2.5}$$

and
$$\mathrm{E}\left(\frac{\partial^2 l}{\partial \theta^2}\right) + \mathrm{E}\left(\frac{\partial l}{\partial \theta}\right)^2 = 0. \tag{2.6}$$

We have from (2.4) that

$$l = [y\theta - b(\theta)]/a(\phi) + c(y, \phi),$$

whence
$$\frac{\partial l}{\partial \theta} = [y - b'(\theta)]/a(\phi) \tag{2.7}$$

and
$$\frac{\partial^2 l}{\partial \theta^2} = -b''(\theta)/a(\phi). \tag{2.8}$$

Thus from (2.5) and (2.7) we have

$$0 = \mathrm{E}\left(\frac{\partial l}{\partial \theta}\right) = [\mu - b'(\theta)]/a(\phi),$$

and so

$$\mathrm{E}(Y) = \mu = b'(\theta).$$

Similarly from (2.6), (2.7) and (2.8) we have

$$0 = -\frac{b''(\theta)}{a(\phi)} + \frac{\operatorname{var}(Y)}{a^2(\phi)},$$

so that

$$\operatorname{var}(Y) = b''(\theta)a(\phi),$$

where primes denote differentiation with respect to θ. Thus the variance of Y is the product of two functions; one, $b''(\theta)$, depends on the canonical parameter (and hence on the mean) only and will be called the *variance function*, while the other is independent of θ and depends only on ϕ.

The function $a(\phi)$ is commonly of the form

$$a(\phi) = \phi/w,$$

where w is a *prior weight*, assumed known, and ϕ will be called the *dispersion parameter* (sometimes denoted by σ^2). Thus for a Normal model in which each observation is the mean of n independent readings we have

$$a(\phi) = \sigma^2/n,$$

so that $w = n$.

The most important distributions of this type with which we shall be concerned are summarized in Table 2.1.

Table 2.1 *Characteristics of some distributions of the exponential family*

	Normal	*Poisson*	*Binomial*	*Gamma*	*Inverse Gaussian*
Range of y	$(-\infty, \infty)$	$0(1)\infty$	$\frac{0(1)n}{n}$	$(0, \infty)$	$(0, \infty)$
$a(\ )$	$\phi = \sigma^2$	1	$1/n$	$\phi = \nu^{-1}$	$\phi = \sigma^2$
$b(\ )$	$\frac{1}{2}\theta^2$	e^θ	$\ln(1+e^\theta)$	$-\ln(-\theta)$	$-(-2\theta)^{\frac{1}{2}}$
$c(\ )$	$-\frac{1}{2}\left(\frac{y^2}{\phi}+\ln(2\pi\phi)\right)$	$-\ln y!$	$\ln\left[\binom{n}{ny}\right]$	$(\nu-1)\ln(y\nu)+\ln\nu-\ln\Gamma(\nu)$	$-\frac{1}{2}\left(\ln(2\pi\phi y^3)+\frac{1}{\phi y}\right)$
$\mu = E(Y)$	θ	e^θ	$e^\theta/(1+e^\theta)$	$-1/\theta$	$(-2\theta)^{-\frac{1}{2}}$
Variance function	1	μ	$\mu(1-\mu)$	μ^2	μ^3

2.2.3 *Link functions*

The link function relates the linear predictor η to the expected value μ of a datum y. In classical linear models they are identical, and the identity link is 'sensible' in the sense that both η and μ can take any value on the real line. However, when we are dealing with counts and the distribution is Poisson, we must have $\mu > 0$, so that the identity link is less attractive in part because η may then be negative. Models based on independence of probabilities associated with the different classifications of cross-classified data lead naturally to considering multiplicative effects, and this is expressed by the log link, $\eta = \log \mu$ with its inverse $\mu = e^{\eta}$. Now additive effects contributing to η become multiplicative effects contributing to μ.

For the binomial distribution we have $0 < \mu < 1$ and a link should satisfy the condition that it maps the interval (0, 1) onto the whole real line. We shall consider three such functions in subsequent chapters, namely:

1. logit
$$\eta = \log[\mu/(1-\mu)];$$
2. probit
$$\eta = \Phi^{-1}(\mu)$$
where $\Phi(\ )$ is the Normal cumulative distribution function;
3. complementary log–log
$$\eta = \log[-\log(1-\mu)].$$

The power family of links is important, and can be specified either by

$$\eta = (\mu^{\alpha} - 1)/\alpha, \tag{2.9a}$$

with the limiting value

$$\eta = \log \mu; \qquad \text{as } \alpha \to 0, \tag{2.9b}$$

or by

$$\begin{aligned} \eta &= \mu^{\alpha}; & \alpha &\neq 0, \\ \eta &= \log \mu; & \alpha &= 0. \end{aligned} \tag{2.10}$$

The first form has the advantage of a smooth transition as α passes through zero, but in both forms special action has to be taken in any computation with $\alpha = 0$.

2.2.4 *Sufficient statistics and canonical links*

Each of the distributions in Table 2.1 has a special link function for which there exist sufficient statistics for the parameters in the linear predictor $\eta = \Sigma\beta_i x_i$. These *canonical links*, as they will be called, occur when

$$\theta = \eta,$$

where θ is the canonical parameter, as shown in Table 2.1. The canonical links for the distributions in Table 2.1 are thus:

Normal	$\eta = \mu$,
Poisson	$\eta = \ln \mu$,
binomial	$\eta = \ln[\mu/(1-\mu)]$,
gamma	$\eta = \mu^{-1}$,
inverse Gaussian	$\eta = \mu^{-2}$.

For the canonical links, the sufficient statistics are given by $\Sigma Y x_j$, $j = 1, \ldots, p$, summation being over the units. It should be stressed, however, that, although the canonical links lead to desirable statistical properties of the model, particularly in small samples, there is in general no *a priori* reason why the systematic effects in the model should be additive on the scale given by that link. It is convenient if they are, but no more, and in later chapters we shall deal with several models in which non-canonical links are used.

2.3 Measuring the goodness of fit

2.3.1 *The discrepancy of a fit*

Fitting a model to data may be regarded as a way of replacing a set of data values $\mathbf{y}$ by a set of fitted values $\hat{\boldsymbol{\mu}}$ derived from a model involving (usually) a relatively small number of parameters. In general the μs will not equal the ys exactly, and the question then arises of how discrepant they are, because while a small discrepancy may be tolerable a large discrepancy is not. Measures of discrepancy (or goodness of fit) may be formed in various ways, but we shall be primarily concerned with that formed from the logarithm of a ratio of likelihoods, to be called the *deviance*.

Given N observations we can fit models to them containing up to

N parameters. The simplest model, the *null model*, has one parameter, representing a common μ for all the ys; the null model consigns all the variation between the ys to the random component. At the other extreme the *full model* has N parameters, one per observation, and the μs derived from it match the data exactly. The full model consigns all the variation in the ys to the systematic component leaving none for the random component.

In practice the null model is too simple and the full model is uninformative because it does not summarize the data but merely repeats them in full. However, the full model gives us a baseline for measuring the discrepancy for an intermediate model with p parameters.

It is convenient to express the log likelihood in terms of the mean-value parameter $\boldsymbol{\mu}$ rather than the canonical parameter $\boldsymbol{\theta}$. Let $l(\hat{\boldsymbol{\mu}}, \phi; \mathbf{y})$ be the log likelihood maximized over $\boldsymbol{\beta}$ but not over the nuisance parameter ϕ. The maximum likelihood achievable in a full model with N parameters is $l(\mathbf{y}, \phi; \mathbf{y})$ which is generally finite. The discrepancy of a fit is proportional to twice the difference between the maximum log likelihood achievable and that achieved by the model under investigation. If we denote by $\hat{\boldsymbol{\theta}} = \boldsymbol{\theta}(\hat{\boldsymbol{\mu}})$ and $\tilde{\boldsymbol{\theta}} = \boldsymbol{\theta}(\mathbf{y})$ the estimates of the canonical parameters under the two models, the discrepancy, assuming $a_i(\phi) = \phi/w_i$, can be written

$$\Sigma\, 2w_i[y_i(\tilde{\theta}_i - \hat{\theta}_i) - b(\tilde{\theta}_i) + b(\hat{\theta}_i)]/\phi = D(\mathbf{y}; \hat{\boldsymbol{\mu}})/\phi,$$

where $D(\mathbf{y}; \hat{\boldsymbol{\mu}})$ is known as the deviance for the current model and is a function of the data only.

The forms of the deviances for the distributions given in Table 2.1 are as follows, summation being over $i = 1, \ldots, N$:

Normal	$\Sigma(y-\hat{\mu})^2$,
Poisson	$2\{\Sigma[y \ln(y/\hat{\mu}) - (y-\hat{\mu})]\}$,
binomial	$2\Sigma\{[y \ln(y/\hat{\mu})] + (n-y)\ln[(n-y)/(n-\hat{\mu})]\}$,
gamma	$2\Sigma[-\ln(y/\hat{\mu}) + (y-\hat{\mu})/\hat{\mu}]$,
inverse Gaussian	$\Sigma(y-\hat{\mu})^2/(\hat{\mu}^2 y)$.

For the Normal distribution the deviance is just the residual sum of squares, while for the Poisson it is the statistic labelled G^2 by Bishop *et al.* (1975) and others. Note that the second terms in the expressions for the Poisson and gamma deviances are usually identically zero (for details, see Nelder and Wedderburn, 1972).

The other important measure of discrepancy is the generalized Pearson X^2 statistic, which takes the form

$$X^2 = \Sigma\ (y-\hat{\mu})^2/V(\hat{\mu}),$$

where $V(\hat{\mu})$ is the estimated variance function for the distribution concerned. For the Normal distribution, X^2 is again the residual sum of squares, while for the Poisson or binomial distributions it is the original Pearson X^2 statistic.

Both the deviance and the generalized Pearson X^2 have exact χ^2 distributions for Normal-theory linear models (assuming of course that the model is true), and asymptotic results are available for the other distributions. However, asymptotic results may not be specially relevant to limited amounts of data, and for these either D or X^2 may prove superior in its distributional properties. The deviance has a general advantage as a measure of discrepancy in that it is additive for nested sets of models if maximum-likelihood estimates are used, whereas X^2 in general is not. However, X^2 might be preferred in some circumstances because of its more direct interpretation.

2.3.2 *The analysis of deviance*

The analysis of variance, particularly when applied to orthogonal data with Normal errors, is a highly useful technique for screening the effects of factors and their interactions. We need some generalization of it that would be applicable to the wider class of generalized linear models. There are two problems in the generalization that need consideration: first, the terms in the model will, in general, no longer be orthogonal and, secondly, sums of squares will, for non-Normal distributions, no longer be appropriate measures of the contribution of a term to the total discrepancy. The second problem is the more easily dealt with, and we consider it first. The terms in the analysis of variance can usefully be thought of as the first differences of the goodness-of-fit statistic for a sequence of models, each including one term more than the previous one. Thus the factorial model for two factors A and B gives rise to an analysis of variance with three terms A, B and $A \cdot B$. The sums of squares for these are the first differences of the residual sums of squares obtained from fitting successively the models 1, A, $A+B$ and $A+B+A\cdot B$. As an example, consider the following analysis of an unreplicated 4×3 table indexed by A and B:

			Analysis of variance		
Model	*d.f.*	*Discrepancy*	*s.s.*	*d.f.*	*Term*
1	11	1000			
			400	3	A ignoring B
A	8	600			
			200	2	B eliminating A
$A+B$	6	400			
			400	6	$A\cdot B$ eliminating A and B
$A+B+A\cdot B$	0	0			

On the left is the sequence of models with their discrepancies, as measured by the residual sums of squares; note that the last model is the full model, i.e. has as many parameters as observations, so that the degrees of freedom (d.f.) and the discrepancy are both zero. On the right is the analysis-of-variance (anova) table, with the sums of squares (s.s.) obtained from the first differences of the discrepancies. Note that the discrepancy for model 1 is just the total sum of squares about the mean in the anova table.

The form of the generalization is now clear. Given a sequence of nested models we can use the deviance as our generalized measure of discrepancy, and form an analysis-of-deviance (anodev) table by taking the first differences, as before. However, the interpretation of the anodev table is now complicated by the non-orthogonality of the terms. Each number represents the variation accounted for by that term having eliminated those terms above it, but ignoring any effects of those terms below it. We may thus need to consider many model sequences, each producing its own analysis-of-deviance table. Note that this problem is present with classical linear models when non-orthogonality occurs. We shall not discuss here the various strategies that have been proposed for generating and looking at the goodness of fit of sets of model sequences. Suffice it to say that the aim of these strategies is to produce parsimonious models for the data in which terms that are not necessary are excluded. Note the use of plural in 'models'; it is most unlikely with complex data that a single model will be a clear winner, and it can be most misleading to quote only the 'best' model, when several others are very close to it in terms of goodness of fit.

Once we depart from the Normal-theory linear model we generally lack an exact theory for the distribution of the deviance. In certain special cases, for example with exponential or inverse Gaussian observations in a simple design, exact results can be found. Usually we rely on the χ^2 approximation for differences between deviances (Section 8.4). In some circumstances the deviance may be approximated by χ^2, for example in discrete data problems where the counts are large. In general, however, the χ^2 approximation for the deviance is not very good even as $N \to \infty$. Further work on the asymptotic distribution of $D(\mathbf{Y}; \hat{\boldsymbol{\mu}})$ remains to be done. In the meantime the analysis-of-deviance table should be regarded as a screening device for picking out obviously important terms, and no attempt should be made to assign significance levels to the raw deviances.

2.4 Residuals

For Normal models we can express the dependent variate in the form

$$y = \hat{\mu} + (y - \hat{\mu}),$$

i.e. datum = fitted value + residual. Residuals can be used to explore the adequacy of fit of a model, in respect of both choice of variance function and terms in the linear predictor. Residuals may also indicate the presence of anomalous values, which require further investigation (see Chapter 11). For generalized linear models we require a generalization of residuals, applicable to all the distributions which may replace the Normal, and which can be used for the same purposes as the standard Normal residuals. We shall use the theoretical form, involving μ rather than $\hat{\mu}$, and we define three forms of generalized residual, which we call the Pearson, Anscombe and deviance residuals.

2.4.1 *Pearson residual*

The Pearson residual

$$r_{\mathrm{P}} = \frac{y - \mu}{\sqrt{V(\mu)}} \tag{2.11}$$

is just the simple residual scaled by the estimated standard deviation of Y. The name is taken from the fact that for the Poisson distribution the Pearson residual is just the signed square root of the component of the Pearson X^2 goodness-of-fit statistic, so that

$$\Sigma r_{\mathrm{P}}^2 = X^2.$$

2.4.2 *Anscombe residual*

A disadvantage of the Pearson residual is that the distribution of r_{P} for non-Normal distributions will often be markedly skewed, and so may fail to have properties similar to those of the Normal residual. Anscombe proposed defining a residual using a function $A(y)$ in place of y, where $A(\ )$ is chosen to make the distribution of $A(Y)$ 'as Normal as possible'. Wedderburn (unpublished, but see Barndorff-Nielsen, 1978) showed that, for the likelihood functions occurring in generalized linear models, the function $A(\ \ )$ is given by

$$A(\ \ ) = \int \frac{\mathrm{d}\mu}{V^{\frac{1}{3}}(\mu)}.$$

Thus for the Poisson distribution we have

$$\int \frac{\mathrm{d}\mu}{\mu^{\frac{1}{3}}} = \tfrac{3}{2}\mu^{\frac{2}{3}},$$

so that we base our residual on $y^{\frac{2}{3}} - \mu^{\frac{2}{3}}$. Now the transformation that 'Normalizes' the probability function does not stabilize the variance, so that we must scale by dividing by the square root of the variance of $A(Y)$, which is, to the first order, $A'(\mu)\sqrt{V(\mu)}$. Thus for the Poisson distribution the Anscombe residual, to be denoted by r_{A}, is given by

$$r_{\mathrm{A}} = \frac{\frac{3}{2}(y^{\frac{2}{3}} - \mu^{\frac{2}{3}})}{\mu^{\frac{1}{6}}}.$$

See Anscombe (1953) and Cox and Snell (1968) for the corresponding residual for the binomial distribution. For the gamma distribution the Anscombe residual takes the form

$$r_{\mathrm{A}} = \frac{3(y^{\frac{1}{3}} - \mu^{\frac{1}{3}})}{\mu^{\frac{1}{3}}}.$$

This cube-root transformation was used by Wilson and Hilferty (1931) to normalize variables with a χ^2 distribution. Similarly the inverse Gaussian distribution gives

$$r_{\mathrm{A}} = (\ln y - \ln \mu)/\mu.$$

2.4.3 *Deviance residual*

If the deviance is used as a measure of discrepancy of a generalized linear model, then each unit contributes a quantity d_i to that measure, so that $\Sigma d_i = D$. Hence if we define

$$r_D = \text{sgn}\,(y-\mu)\sqrt{d_i},$$

we have a quantity which increases (or decreases) with $y_i - \mu_i$ and for which $\Sigma r_D^2 = D$. Thus for the Poisson distribution we have

$$r_D = \text{sgn}\,(y-\mu)\,[2(y\ln(y/\mu)-y+\mu)]^{\frac{1}{2}}.$$

Although the Anscombe and deviance residuals appear to have very different functional forms for non-Normal distributions, the values that they take for given y and μ are often remarkably similar, as is clear from a Taylor series expansion. Consider again the Poisson distribution and set $y = c\mu$, so that

$$r_A = \tfrac{3}{2}\mu^{\frac{1}{6}}(c^{\frac{2}{3}}-1)$$

and
$$r_D = \text{sgn}\,(c-1)\mu^{\frac{1}{2}}[2(c\ln c-c+1)]^{\frac{1}{2}}.$$

If we compare the two functions $\frac{3}{2}(c^{\frac{2}{3}}-1)$ and $[2(c\ln c-c+1)]^{\frac{1}{2}}$ for a range of values of c, we find a close similarity as Table 2.2 (which also shows the Pearson residual) indicates.

Table 2.2 *Comparison of Poisson residuals*

	r_A	r_D	r_P
c	$\frac{3}{2}(c^{\frac{2}{3}}-1)$	$[2(c\ln c-c+1)]^{\frac{1}{2}}$	$c-1$
0	−1.5	−1.414	−1.0
0.2	−0.987	−0.956	−0.8
0.4	−0.686	−0.683	−0.6
0.6	−0.433	−0.432	−0.4
0.8	−0.207	−0.207	−0.2
1.0	0	0	0
1.5	0.466	0.465	0.5
2.0	0.881	0.879	1.0
2.5	1.263	1.258	1.5
3.0	1.620	1.610	2.0
4.0	2.280	2.256	3.0
5.0	2.886	2.845	4.0
10	5.462	5.296	9.0

Within this range the maximum difference between r_A and r_D is about 6% at $c = 0$, and much less over most of the range. The Pearson residual is considerably greater in the upper part of the range but goes less far in the negative direction.

2.5 An algorithm for fitting generalized linear models

We shall show that the maximum-likelihood estimates of the parameters $\boldsymbol{\beta}$ in the linear predictor η can be obtained by iterative weighted least squares. In this regression the dependent variable is not y but z, a linearized form of the link function applied to y, and the weights are functions of the fitted values $\hat{\mu}$. The process is iterative because both the *adjusted dependent variable* z and the weight W depend on the fitted values, for which only current estimates are available. The procedure underlying the iteration is as follows. Let $\hat{\boldsymbol{\eta}}_0$ be the current estimate of the linear predictor, with corresponding fitted value $\hat{\boldsymbol{\mu}}_0$ derived from the link function $\eta = g(\mu)$. Form the adjusted dependent variate with typical value

$$z_0 = \hat{\eta}_0 + (y - \hat{\mu}_0)\left(\frac{d\eta}{d\mu}\right)_0,$$

where the derivative of the link is evaluated at $\hat{\mu}_0$, and the (quadratic) weight defined by

$$W_0^{-1} = \left(\frac{d\eta}{d\mu}\right)_0^2 V_0,$$

where V is the variance function of y. Now regress z_0 on $x_1, x_2, \ldots, x_p$ with weight W_0 to give new estimates $\mathbf{b}_1$ of the parameters; from these form a new estimate $\hat{\boldsymbol{\eta}}_1$, of the linear predictor. Repeat until changes are sufficiently small. Note that z is just a linearized form of the link function applied to the data, for to the first order

$$g(y) \simeq g(\mu) + (y - \mu)g'(\mu)$$

and the right-hand side is

$$\eta + (y - \mu)\frac{d\eta}{d\mu}.$$

The variance of Z is just W^{-1} (ignoring the dispersion parameter), assuming that η and μ are fixed and known. In this formulation the way in which the calculations for the regression are done is left open; we discuss some possibilities in Section 3.8.

A convenient feature of generalized linear models is that they have a simple starting procedure necessary to allow the iteration to get under way. This consists of using the data themselves as the first estimate of $\hat{\mu}_0$ and from this deriving $\hat{\eta}_0$, $(\mathrm{d}\eta/\mathrm{d}\mu)_0$ and V_0. Adjustments may be required to the data to prevent, for example, our trying to evaluate ln (0) for the log link applied to counts when the datum value is zero. These will be described under the appropriate chapters, as will various complexities sometimes associated with the convergence of the iterative process.

2.5.1 *Justification of the fitting procedure*

From the log likelihood l, written in canonical form

$$l = [y\theta - b(\theta)]/a(\phi) + c(y, \phi),$$

we have

$$\partial l/\partial\theta = [y - b'(\theta)]/a(\phi) = (y-\mu)/a(\phi).$$

Hence

$$\frac{\partial l}{\partial \mu} = \frac{\partial l}{\partial \theta}\bigg/\frac{\mathrm{d}\mu}{\mathrm{d}\theta} = \frac{y-\mu}{V},$$

since

$$\mathrm{d}\mu/\mathrm{d}\theta = b''(\theta) = V/a(\phi).$$

Now

$$\frac{\partial l}{\partial \eta} = \frac{\partial l}{\partial \mu}\frac{\mathrm{d}\mu}{\mathrm{d}\eta},$$

and finally

$$\frac{\partial l}{\partial \beta_i} = \frac{\partial l}{\partial \eta}\frac{\partial \eta}{\partial \beta_i} = \frac{\partial l}{\partial \mu}\frac{\mathrm{d}\mu}{\mathrm{d}\eta}x_i = \left(\frac{y-\mu}{V}\right)\frac{\mathrm{d}\mu}{\mathrm{d}\eta}x_i.$$

The maximum-likelihood equations for β_i are therefore given by

$$\Sigma\left(\frac{y-\mu}{V}\right)\frac{\mathrm{d}\mu}{\mathrm{d}\eta}x_i = \Sigma W(y-\mu)\frac{\mathrm{d}\eta}{\mathrm{d}\mu}x_i = 0$$

for each variate x_i, with summation over the units.

Fisher's scoring method uses the expected value of the Hessian matrix, i.e.

$$\mathrm{E}\left(\frac{\partial^2 l}{\partial\beta_i\,\partial\beta_j}\right) = \mathrm{E}\left(\frac{\partial}{\partial\beta_j}\left[\Sigma\,(Y-\mu)V^{-1}\frac{\mathrm{d}\mu}{\mathrm{d}\eta}x_i\right]\right)$$

$$= \mathrm{E}\left(\Sigma\,(Y-\mu)\frac{\partial}{\partial\beta_j}\left[V^{-1}\frac{\mathrm{d}\mu}{\mathrm{d}\eta}x_i\right] + \Sigma\frac{\partial}{\partial\beta_j}(Y-\mu)V^{-1}\frac{\mathrm{d}\mu}{\mathrm{d}\eta}x_i\right) \tag{2.12}$$

$$= -\sum V^{-1}\left(\frac{d\mu}{d\eta}\right)^2 x_i x_j$$

$$= -\sum W x_i x_j. \tag{2.13}$$

Thus given current estimates **b** of $\boldsymbol{\beta}$, the method gives adjustments $\delta\mathbf{b}$ defined by

$$\mathbf{A}\delta\mathbf{b} = \mathbf{c},$$

where **A** is a $p \times p$ matrix given by

$$A_{ij} = \sum_k W_k x_{ik} x_{jk}$$

and **c** is a $p \times 1$ vector given by

$$c_i = \sum_k W_k x_{ik}(y_k - \mu_k)\, d\eta_k/d\mu_k.$$

Now

$$(\mathbf{Ab})_i = \sum_j A_{ij} b_j = \sum_k W_k x_{ik} \eta_k,$$

and therefore new estimates $\mathbf{b}^* = \mathbf{b} + \delta\mathbf{b}$ satisfy the equations

$$(\mathbf{Ab}^*)_i = [\mathbf{A}(\mathbf{b}+\delta\mathbf{b})]_i = \sum W_k x_{ik}[\eta_k + (y_k - \mu_k)\, d\eta_k/d\mu_k],$$

and these have the form of linear weighted least-squares equations with weight

$$W = V^{-1}\left(\frac{d\mu}{d\eta}\right)^2$$

and dependent variate

$$z = \eta + (y - \mu)\frac{d\eta}{d\mu}.$$

Note that simplification occurs for the canonical links; for these the expected value and the actual value of the Hessian matrix coincide, so that the Fisher scoring method and the Newton–Raphson method reduce to the same algorithm. This comes about because the linear weight function $V^{-1}\,d\mu/d\eta$ in the maximum-likelihood equations is a constant, and the first term in the expansion of the

Hessian (2.12) is identically zero. Note also that $W = V$ for this case. Finally, if the model is linear on the scale on which Fisher's information is constant, i.e. $g'(\mu) = V^{-\frac{1}{2}}(\mu)$, the vector of weights is constant and need not be recomputed at each iteration.

CHAPTER 3

Models for continuous data with constant variance

3.1 Introduction

Generalized linear models are essentially a generalization of classical linear models, and this chapter will present these latter models in a framework which makes the generalization a natural one. There is an enormous literature on classical linear models, not all of it helpful to the reader, and no attempt will be made to summarize it here. Draper and Smith (1981) and Seber (1977) are excellent general reference books.

We shall define the models of this chapter in the following form:

$Y_i \sim \mathrm{N}(\mu_i, \sigma^2)$,	$\boldsymbol{\mu} = \boldsymbol{\eta}$,	$\boldsymbol{\eta} = \sum_1^p \beta_j \mathbf{x}_j$,	
errors Normally distributed and independent;	identity link;	linear predictor based on covariates $\mathbf{x}_1, \ldots, \mathbf{x}_p$.	(3.1)

The data vector $\mathbf{y}$, the mean vector $\boldsymbol{\mu}$ and the linear predictor are all of length N. The left-hand component in (3.1) is a specification of the random part of the model. The right-hand components describe the systematic parts of the model including the construction of the linear predictor η from the covariates and the link between η and μ. By suppressing the link, regarding the $\mathbf{x}_i$ as columns of a matrix $\mathbf{X}$, and writing the β_i in vector form, we recover the standard matrix formulation

$$\mathrm{E}(\mathbf{Y}) = \mathbf{X}\boldsymbol{\beta}.$$

Note that we are restricting ourselves to the subclass of linear models where (a) there is only one error component and (b) the errors are assumed independent. We now consider in more detail the random and systematic components of the model.

3.2 Error structure

The error component of classical linear models is defined by the Normal (or Gaussian) distribution. The observations $\mathbf{y}$ are supposed to have means and variances given by

$$E(\mathbf{Y}) = \boldsymbol{\mu}, \qquad \operatorname{cov}(\mathbf{Y}) = \sigma^2\mathbf{I}, \tag{3.2}$$

so that the Ys have equal variances and are uncorrelated.

The assumption of Normality, although important as the basis of an exact small-sample theory, need not be relied upon in large samples: the central limit theorem offers protection from all but the most extreme distributional deviations but there is some loss of efficiency (Cox and Hinkley, 1968). The theory of least squares can be developed using only the first- and second-moment assumptions of (3.2), and without requiring the additional assumption of Normality; this idea is extended in Chapter 8 to cover models where the variance is not constant but varies with μ. The important aspect of (3.2) from the present viewpoint is that the variance is the same for all values of μ. This assumption requires checking – usually by an examination of residuals (Chapter 11).

The frequency function of the Normal distribution takes the form

$$\frac{1}{\sqrt{(2\pi\sigma^2)}}\exp\left(-\frac{(y-\mu)^2}{2\sigma^2}\right); \qquad -\infty < y < \infty,$$

and its shape is shown in Fig. 3.1. The distribution is symmetrical with mode, mean and median all at μ, and σ (the standard deviation) is the horizontal distance between the mean and the point of inflection. About 68%, 95% and 99.8% of the distribution lies in the range $\pm\sigma$, $\pm 2\sigma$ and $\pm 3\sigma$ of the mean, respectively. The log-likelihood function for a single observation with known variance is a parabola, with a maximum at y and constant second derivative $-1/\sigma^2$.

The Normal distribution is relevant primarily to measurements of continuous quantities, though it is sometimes used as an approximation for discrete measurements. It is frequently used to model data, such as weights, which though continuous are essentially positive, although the distribution itself covers the whole real line. Such a use is acceptable in practice provided that the data values are sufficiently

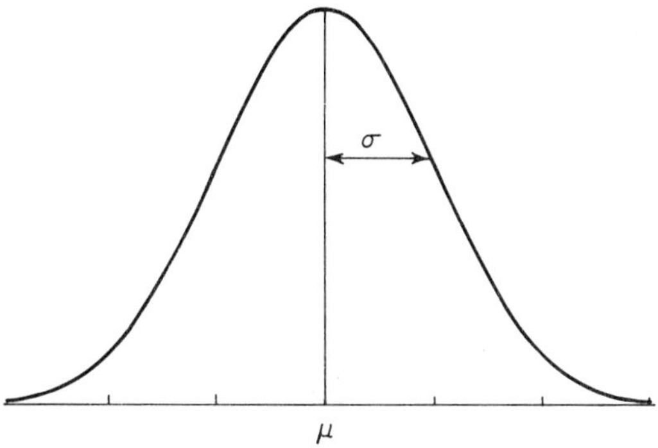

Fig. 3.1. *The Normal (or Gaussian) distribution with mean μ and standard deviation σ.*

far away from zero. If, for example, data have a mean of 100 and a standard deviation of 10, the part of the corresponding Normal distribution in the negative half of the real line is negligible practically. If data y which are essentially positive approach the zero point at all closely then it will usually be found that the data themselves contradict the assumption of constant variance independent of the mean. The use of $\log y$ in place of y may then justify the assumption of Normal errors with constant variance. (See Chapter 7 for models using another distribution, the gamma distribution, for which observations must be positive.)

3.3 Systematic component (linear predictor)

The expected value μ of a datum y is identical to the linear predictor η in these models and η is assumed to be expressible in the form

$$\eta = \sum_{1}^{p} \beta_j x_j,$$

where $x_1, x_2, \ldots, x_p$ are *covariates* with associated *parameters* β_i. The terms 'stimulus variables', 'explanatory variables', 'independent variables', and others are frequently used for the x_j, whose values we shall assume are known effectively without error. The covariates can be of several types, which we now discuss. It is important to note that the material in Sections 3.3 to 3.5 is relevant to *all* generalized linear models.

3.3.1 *Continuous covariates*

These comprise covariates such as mass, temperature, amount of fertilizer, concentration of a solute, etc., which can take values on a continuous scale. Models containing only continuous covariates are often termed *regression models* and are contrasted with analysis-of-variance (anova) models involving discrete covariates. We shall not make this distinction, and indeed by introducing mixed terms in Section 3.3.4 we shall deliberately seek to blur it in order to produce a more general class of models. A continuous covariate x may be rescaled quite arbitrarily to produce a function $f(x)$ without destroying the linearity of the model, provided that $f(x)$ does not contain unknown parameters that enter non-linearly. In particular we may use $x^2, x^3, \ldots$ as well as x to build up a polynomial in x without destroying the linearity. Similarly the terms $\beta_0+\beta_1 x_1+\beta_2 x_2$ may be expanded to include the product term $\beta_{12} x_1 x_2$, producing a bilinear relationship. If these terms are rearranged in the form

$$(\beta_0+\beta_2 x_2)+(\beta_1+\beta_{12} x_2)x_1,$$

they can be seen as expressing a linear relationship with x_1 in which both intercept and slope are linear functions of x_2. The rearrangement

$$(\beta_0+\beta_1 x_1)+(\beta_2+\beta_{12} x_1)x_2$$

expresses the complementary form.

A covariate such as $\exp(\gamma x)$, however, will produce a non-linearity unless γ is known *a priori*; if γ is not known the model can be fitted as linear for any fixed (but arbitrary) γ, and the optimum value of γ can be determined as that yielding the smallest discrepancy. This leads into the general area of models that are partly linear and partly non-linear; these will be discussed in Chapter 10.

3.3.2 *Qualitative covariates*

The linear predictor may include terms of the form $\alpha_i, i = 1, \ldots, k$, where i indexes a *classifying factor* which divides the data into k disjunct groups. The groups are then assumed to contribute distinct quantities α_i to the linear predictor. Consider, for example, such a factor A together with a continuous covariate x and model

$$\eta_i = \alpha_i+\beta x.$$

Here the slope β is assumed independent of the grouping factor but the intercept varies with it. *Factors*, as we shall call qualitative covariates, occur in many contexts: in designed experiments, for instance, factors may define the block number in which a given experimental unit (plot) lies, or may constitute a treatment factor such as variety in a variety trial; in a survey, factors might include region, sex, marital status, and so on.

A factor can take only a limited set of values, to be called *levels*. A factor with k levels can always be coded using the integers $1, 2, \ldots, k$ (sometimes the coding $0, 1, \ldots, k-1$ is more convenient). We shall call such coding *formal levels*; in practice the levels may have names or numerical values and these we call *actual levels*. Actual levels may have (i) no ordering associated with them, for example the varietal names in a variety trial; or (ii) an ordering but no relative magnitudes for the levels, as in a preference scale; or (iii) an underlying scale, as when the levels are amounts of fertilizer applied in an agricultural experiment. Note that a type (iii) factor might be treated as a continuous covariate whose values happen to be restricted to a few values by the experimental design; if we treat it as a factor we fit a separate effect for each level in a quite unstructured way, whereas if we treat it as a variate we impose a particular form of response (i.e. a linear one) on the model.

Frequently data are cross-classified by many factors simultaneously. If A, B and C are three such factors with indices i, j and k respectively, then the simplest model we can define would include terms of the form

$$\alpha_i + \beta_j + \gamma_k,$$

the so-called 'main effects'. This model implies that if we look at 'slices' of the data for each level of A we shall find that they can be modelled by effects of B and C that are additive and the same in each slice. Similarly for the other factors. It may be necessary, however, to include terms analogous to $\beta_{12} x_1 x_2$ with continuous covariates. Such terms, of the form $(\alpha\beta)_{ij}$, implying a separate effect for each combination of indices i and j in the two-way array classified by A and B, comprise *interactions*. We shall refer to a term such as $(\alpha\beta)_{ij}$ as a two-factor interaction, but the term 'first-order interaction' is often used, the order being one less than the number of factors involved. The relation between interactions and main effects has been the source of much confusion in the literature and we shall consider it in more detail in Section 3.5.

3.3.3 *Dummy variates*

The term α_i in a model, where i is the index of factor A with k levels, can be written

$$\alpha_1 \mathbf{u}_1 + \alpha_2 \mathbf{u}_2 + \ldots + \alpha_k \mathbf{u}_k,$$

where the $\mathbf{u}_j$ are *dummy variates* taking the value 1 if the unit has factor A at level j, and zero elsewhere. Thus if $k = 3$ and the formal levels for five units are 1 2 2 3 3, the dummy variates $(\mathbf{u}_1, \mathbf{u}_2, \mathbf{u}_3)$ take values as follows:

Unit	*A*	$\mathbf{u}_1$	$\mathbf{u}_2$	$\mathbf{u}_3$
1	1	1	0	0
2	2	0	1	0
3	2	0	1	0
4	3	0	0	1
5	3	0	0	1

A compound term such as $(\alpha\beta)_{ij}$ has dummy variates whose values are products of corresponding elements of $\mathbf{u}_i$ and $\mathbf{v}_j$, the dummy variates for A and B as single-factor terms. Note that in the example above

$$\mathbf{u}_1 + \mathbf{u}_2 + \mathbf{u}_3 \equiv \mathbf{1}$$

irrespective of the allocation of levels to units. Now $\mathbf{1}$ is the dummy variate corresponding to the intercept term (often written μ) in the linear predictor. This relation between the terms μ and α_i is a simple example of *intrinsic aliasing* which will be discussed in Section 3.5.

3.3.4 *Mixed terms*

In Section 3.3.2 we considered a model

$$\eta_i = \alpha_i + \beta x,$$

where the intercept varied with the level of a factor but the slope did not. However, sometimes the slope may also change with the level, requiring the term βx to be replaced by the more general $\beta_i x$. Terms in the linear predictor where a slope parameter changes with the level of one or more factors we call *mixed terms*, as they include aspects of both continuous and qualitative covariates. It is important that any computer program for linear models should allow mixed terms to be specified as easily as continuous or qualitative terms, because the assumption frequently made, that a slope is the same for all levels of some classifying factor, ought to be able to be tested during the process of model fitting by allowing it to vary.

Dummy variates for mixed terms take the same general form as those for factors except that the 1s are replaced by the corresponding x-values. Using the same factor values as in Section 3.3.3 and with a variate x as shown, the dummy variates ($\mathbf{u}_1, \mathbf{u}_2, \mathbf{u}_3$) for the mixed term $\beta_i x$ take values as follows:

Unit	*A*	*x*	$\mathbf{u}_1$	$\mathbf{u}_2$	$\mathbf{u}_3$
1	1	1	1	0	0
2	2	3	0	3	0
3	2	5	0	5	0
4	3	7	0	0	7
5	3	9	0	0	9

3.4 Model formulae for linear predictors

We now introduce a useful notation for the specification of linear predictors in generalized linear models, a notation which is both compact and easily adapted for use in computer programs (Wilkinson and Rogers, 1973). The convention is continued that any names beginning with letters from the first half of the alphabet refer to factors, and from the second half to continuous covariates. Indices associated with $A, B, C, \ldots$ are $i, j, k, \ldots$. Basic terms in a linear predictor, in algebraical form and as in a model formula, are as follows (note the use of λ instead of β for the coefficient of the continuous covariate to avoid confusion with the parameters for factor B):

Type of term	*Algebraic*	*Model formula*
Continuous covariate	λx	X
Factor	α_i	A
Mixed	$\lambda_i x$	$A \cdot X$
Compound	$(\alpha\beta)_{ij}$	$A \cdot B$
Compound mixed	$\lambda_{ij} x$	$A \cdot B \cdot X$

In the model-formula version X stands for itself, a single vector, while A stands for its set of dummy variates, one for each level. The remaining types of term similarly stand for the appropriate set of dummy variates. Thus the terms in the model formula represent vector subspaces and do not involve the parameters explicitly. Parameters occur implicitly, one per basis vector in each subspace.

3.4.1 *The dot operator*

This operator, already exemplified in the formation of compound terms, implies the formation of all the direct product vectors of the constituent elements. Note, however, that when A has k levels, $A \cdot A$ leads to k vectors equal to the dummy variates of A and $k(k-1)$ null vectors with all elements zero. Such null vectors contribute nothing to the linear predictor and so we shall omit them, thus making

$$A \cdot A = A.$$

However, in general,

$$X \cdot X \neq X,$$

the left-hand side being a vector with components x^2. (Note that in both the computer programs GLIM and Genstat, where this notation is used, compound terms involving more than one continuous covariate are not allowed in model formulae.) The dot operator is commutative, so that

$$A \cdot B \equiv B \cdot A,$$

and associative, so that

$$(A \cdot B) \cdot C \equiv A \cdot (B \cdot C).$$

3.4.2 *The + operator*

Terms in a model formula can be joined using the operator +, exactly as in the algebraic form. Repetitions of terms are ignored so that

$$A + A \equiv A,$$

it being pointless to specify the same vector subspace twice. The + operator has lower priority than the dot, so that

$$A \cdot B + C \equiv (A \cdot B) + C.$$

The dot is distributive with respect to + so that

$$A\cdot(B+C) \equiv A\cdot B+A\cdot C.$$

3.4.3 *The crossing* ($*$) *and nesting* (/) *operators*

The crossing operator, denoted by $*$, is used mainly to simplify the specification of factorial models, thus

$$A*B \equiv A+B+A\cdot B,$$
$$A*B*C \equiv A+B+C+A\cdot B+A\cdot C+B\cdot C+A\cdot B\cdot C,$$

and so on. In these expansions A and B may themselves be replaced by model formulae. The operator $*$ has higher priority than + but lower priority than dot, thus

$$A*B+C \equiv A+B+C+A\cdot B,$$
$$A*B\cdot C \equiv A+B\cdot C+A\cdot B\cdot C.$$

Note the convention in expanding expressions that all simple terms come first, then all terms with two components, and so on. This rule, though not essential, is useful for understanding the intrinsic aliasing in models (Section 3.5). The crossing operator is associative, and distributive with respect to +, for

$$\begin{aligned} A*(B+C) &\equiv A+(B+C)+A\cdot(B+C) \\ &\equiv A+B+C+A\cdot B+A\cdot C \\ &\equiv A+B+A\cdot B+A+C+A\cdot C \\ &\equiv A*B+A*C. \end{aligned}$$

When a compound term such as $A\cdot B$ is preceded in an expanded model formula by both constituent terms A and B, it is called the interaction of A and B. The nature of the interaction term will be discussed further in Section 3.5.

The nesting operator relates to an indexing system which in its simplest form has two indices i and j but where there are no links between the pairs (i,j) and (i',j), though there are links between (i,j) and (i,j'). Thus i defines a grouping and j an element within a group. An example would be a set of plant varieties grouped as early, mid-season, or late in cropping; within each group would be distinct varieties, no variety being in more than one group. Two varieties may be connected by being in the same group (i the same) but there is no connection between the first variety in each of two groups (j the same). The appropriate linear predictor for nesting is written as

$$A/B \equiv A + A \cdot B,$$

where / is the nesting operator, and $A \cdot B$ is now preceded by only one constituent term. The interpretation of $A \cdot B$ is now that of B within A. Again A and B may themselves be model formulae with the rule that if pt(A) denotes the product term (using dots) of all elements in A then A/B is defined by

$$A/B \equiv A + \text{pt}(A) \cdot B.$$

Thus, for example,

$$(A*B)/C \equiv A*B + A \cdot B \cdot C.$$

The nesting operator is associative, so that

$$A/(B/C) \equiv (A/B)/C,$$

and distributive with respect to +, since

$$A/(B+C) \equiv A + A \cdot (B+C) \equiv A + A \cdot B + A + A \cdot C \equiv A/B + A/C.$$

Like the crossing operator, the nesting operator has a priority between dot and +, and conventionally we shall give it higher priority than $*$.

3.4.4 *Operators for removal of terms*

The operator $-$ has the obvious meaning for the removal of terms in a model formula. Thus

$$A*B - A \cdot B \equiv A + B.$$

It is sometimes required to remove from a model all the compound terms which include a given factor or factors. The two operators $-/$ and $-*$ cater for this; $-/A$ means 'remove all compound terms which include A (i.e. excluding A itself)' while $-*A$ means 'remove all terms which include A. Thus

$$A*B*C \; -/A \equiv A + B + C + B \cdot C$$

and

$$A*B*C \; -*A \equiv B + C + B \cdot C.$$

3.4.5 *Exponential operator*

If M is a model formula and I is an integer then

$$M{*}{*}I \equiv M{*}M{*}\ldots{*}M,$$

the r.h.s. containing I Ms. The operator is useful for specifying factorial models including terms up to a given level of interaction; thus

$$(A+B+C){*}{*}2 \equiv A+B+C+A\cdot B+A\cdot C+B\cdot C.$$

This operator has highest priority.

We shall use this notation for the specification of linear predictors wherever possible, and readers should bear in mind that the notation (more strictly a subset of it) can be used directly for this purpose in the computer systems Genstat and GLIM.

3.5 Aliasing

Each term in a model formula describes a set of covariates to be included in a linear predictor. If such a set is denoted by $\mathbf{x}_1, \mathbf{x}_2, \ldots, \mathbf{x}_p$, the $\mathbf{x}$s being vectors of length N, the number of units, then the $\mathbf{x}$s can be thought of as defining p directions in an N-dimensional Euclidean space. These p directions define a subspace of up to p dimensions, and this maximum dimensionality will be achieved if no linear relation exists between the $\mathbf{x}$s, i.e. if no quantities $\xi_j, j = 1, \ldots, p$, not all zero, can be found such that

$$\sum_1^p \xi_j \mathbf{x}_j = \mathbf{0}.$$

If k such independent linear relations exist then the $\mathbf{x}$ vectors span a subspace of dimension $p-k$. Generally the individual terms in a model formula will form subspaces of maximum dimensionality, with loss of dimensionality occurring only when we consider the joint subspaces covered by more than one term. The simplest example of loss of dimensionality within a term occurs when a factor level does not occur in any unit, i.e. the corresponding group defined by that level is empty. The dummy vector for that group then has all its elements zero and so defines no direction in the vector space.

We now consider the possible relationships between the subspaces defined by two terms in a model formula. We denote the terms by

P and Q, their dimensionalities by p and q respectively, and assume that $p \geqslant q$. There are three possible types of relationship between P and Q:

1. All the $p+q$ vectors defining P and Q are linearly independent, so that the dimensionality of $P+Q$ is $p+q$.
2. All the vectors of Q are linear combinations of those of P, so that the dimensionality of $P+Q$ is p.
3. There are k independent linear combinations of Q vectors identical to k independent linear combinations of P vectors, $0 < k < q$.

Analogous Venn diagrams for the three cases are shown in Fig. 3.2. Clearly (1) and (2) are extreme cases of (3) for which $k = 0$ and p respectively; however, these cases are usefully distinguished from the general form (3). Note the special case of (3) when $p = q = k$, so that P and Q span identical subspaces.

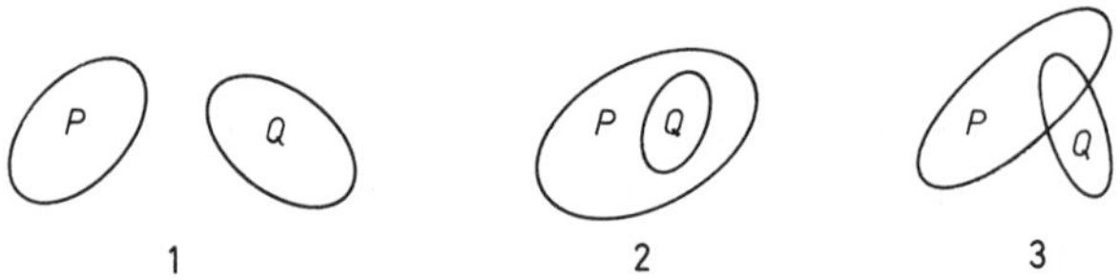

Fig. 3.2. *Venn diagrams for relations between subspaces of terms in a linear model:* 1, *P and Q linearly independent;* 2, *Q entirely aliased with P;* 3, *Q partially aliased with P.*

The effect of overlapping subspaces of the terms in a generalized linear model is to produce *aliasing*, because certain combinations of parameters in the model cannot be distinguished from each other. They have the same estimate for any set of data and there exists no independent information about the components. Consider, for example, measurements on leaves all of the same shape (i.e. in which area = constant × length × breadth), with length and breadth being measured as well as area. Suppose the terms in the model are

$$x_1 = \log \text{length},$$
$$x_2 = \log \text{breadth},$$
$$x_3 = \log \text{area},$$

so that a linear predictor is to be formed as

$$\eta = \beta_0 + \beta_1 x_1 + \beta_2 x_2 + \beta_3 x_3.$$

Now, since area = constant × length × breadth

$$x_3 = c + x_1 + x_2, \tag{3.3}$$

where c is the logarithm of the shape constant. Hence

$$\begin{aligned}\eta &= \beta_0 + \beta_1 x_1 + \beta_2 x_2 + \beta_3(c + x_1 + x_2) \\ &= \beta_0 + \beta_3 c + (\beta_1 + \beta_3)x_1 + (\beta_2 + \beta_3)x_2.\end{aligned}$$

Thus we can distinguish the three quantities

$$\beta_0 + \beta_3 c, \qquad \beta_1 + \beta_3, \qquad \beta_2 + \beta_3,$$

but not the four components β_0, β_1, β_2, β_3 separately. If we write x_0 for the vector of 1s, i.e. the dummy vector for the term β_0, then from (3.3) we have x_3 as a linear combination of x_0, x_1 and x_2, so that the subspace for x_3 (with one dimension) is contained in the sum of the subspaces for x_0, x_1 and x_2. Notice that not only have we assumed that the leaves are of the same shape in the sense that the area is proportional to length × breadth, but we have also assumed that length and breadth show some independent variation between leaves. For if the breadth were proportional to length we should have

$$x_2 = \lambda + x_1,$$

and then the linear predictor would reduce to

$$\begin{aligned}\eta &= \beta_0 + \beta_1 x_1 + \beta_2(\lambda + x_1) + \beta_3(c + x_1 + \lambda + x_1) \\ &= \beta_0 + \beta_2 \lambda + \beta_3(c + \lambda) + (\beta_1 + \beta_2 + 2\beta_3)x_1.\end{aligned}$$

Now only two quantities, namely

$$\beta_0 + \beta_2 \lambda + \beta_3(c + \lambda) \qquad \text{and} \qquad \beta_1 + \beta_2 + 2\beta_3,$$

are distinguishable and the dimensionality of x_0, x_1, x_2 and x_3 is reduced to two.

It is important in this example that the aliasing involved is *intrinsic* to the problem. Given that all the leaves are of the same shape and assuming measurements are made without error, it will occur *whatever* are the particular values of the xs. Such *intrinsic aliasing* occurs most commonly, however, where terms involving factors occur in a model.

3.5.1 *Intrinsic aliasing with factors*

Consider a model formula containing a single factor A together with the grand mean, which we write as

$$1+A,$$

where 1 stands for the dummy vector with all elements 1. Algebraically we have

$$\eta_{ij} = \mu+\alpha_i,$$

where i indexes the groups defined by A and j the units within groups. Now the dummy vectors for the factor A add up to that for μ, because for any unit exactly one of the dummy vectors for A has a 1 and the rest all have 0s. Thus μ is aliased with $\Sigma\alpha_i$, and furthermore it is intrinsically aliased because the relation holds whatever the allocation of units to the groups. The relation of μ to the α_i is not symmetric because μ lies wholely in the space of the α_i but not vice versa. We say that μ is *marginal* to the αs, and in consequence the terms in the model

$$\mu+\alpha_i$$

are *ordered* because of the marginality relation. The linear predictor is clearly unchanged if we add a constant to all the α_i and subtract it from μ. This leaves unchanged the quantities $\mu+\alpha_i$ and also any difference of the α_i. More generally it leaves $\Sigma\lambda_i\alpha_i$ unchanged if $\Sigma\lambda_i = 0$, i.e. any linear combination of the α_i is unchanged if it is orthogonal to the aliased combination $\Sigma 1.\alpha_i$. Quantities unaffected by the aliasing are said to be *estimable*. The quantities μ and α_i separately are not estimable, since the aliasing makes them indistinguishable from $\mu+c$ and α_i-c. The way in which the components of the linear predictor are assigned to estimates m and a_i of μ and α_i is *conventional* and is of no significance in assessing the relevance of the model. The ambiguity can be resolved by imposing a constraint on the components, and this leads to unique values for m and a_i in any model. Constraints are a convenient way of resolving ambiguity but it must be stressed that they do no more than this – they do *not* affect the meaning or interpretation of the model. Thus it does not make sense to set up the hypothesis, for example, that $\mu = 0$ when the α_i are not assumed to be known. Three constraints, chosen from an infinity of possibilities, are as follows:

1. $m = 0$, so that the a_i directly describe the group means;
2. $a_1 = 0$, so that the first group mean is m and $a_2, a_3, \ldots$ measure differences of other group means from the first;
3. $\Sigma a_i = 0$, so that m is the mean of group means.

As an example consider four groups with means 6, 9, 12, 13; then the three constraints produce components in the linear predictor with the values:

	1	2	3
m	0	6	10
a_1	6	0	-4
a_2	9	3	-1
a_3	12	6	2
a_4	13	7	3

3.5.2 *Aliasing in two-way tables*

The aliasing pattern of the linear predictor in a cross-classification of two factors is extremely important, and failure to recognize the conventionality of constraints used to resolve it has led to much confusion in the literature (Nelder, 1977). We are concerned here with the linear model

$$1 + A + B + A \cdot B,$$

or algebraically, with the linear predictor

$$\eta_{ij} = \mu + \alpha_i + \beta_j + \gamma_{ij}.$$

The dummy variables for the four terms show the following relationships

$$\Sigma \alpha_i \equiv \mu,$$

$$\Sigma \beta_j \equiv \mu,$$

$$\sum_j \gamma_{ij} \equiv \alpha_i,$$

$$\sum_i \gamma_{ij} = \beta_j,$$

implying that

$$\sum_{ij} \gamma_{ij} \equiv \mu.$$

Thus

$$\mu \text{ is marginal to } \alpha_i, \beta_j \text{ and } \gamma_{ij},$$
$$\alpha_i \text{ is marginal to } \gamma_{ij},$$
$$\beta_j \text{ is marginal to } \gamma_{ij}.$$

The terms are thus ordered as first μ, then α_i and β_j together, then γ_{ij}. The estimable quantities are the linear predictor

$$\eta_{ij} = \mu+\alpha_i+\beta_j+\gamma_{ij},$$
$$\Sigma\lambda_i(\alpha_i+\gamma_{i.}); \quad \text{with } \Sigma\lambda_i = 0,$$
$$\Sigma\lambda_j(\beta_j+\gamma_{.j}); \quad \text{with } \Sigma\lambda_j = 0,$$
$$\text{and } \Sigma\lambda_{ij}\gamma_{ij}; \quad \text{with } \sum_i \lambda_{ij} = \sum_j \lambda_{ij} = 0.$$

where the dot subscript denotes the average over the corresponding index.

The ambiguities about the values of the estimates of individual components can be resolved again by suitable constraints, and we give two examples for a 2×2 table.

The full parameterization has nine components ($\mu, \alpha_1, \alpha_2, \beta_1, \beta_2, \gamma_{11}, \gamma_{12}, \gamma_{21}, \gamma_{22}$) but only four estimable quantities. Thus five constraints must be imposed on the estimates to produce uniqueness. The first method we may call that of symmetric constraints, and is given by

$$\begin{aligned} a_1+a_2 &= 0, \\ b_1+b_2 &= 0, \\ c_{11}+c_{12} &= 0, \\ c_{21}+c_{22} &= 0, \\ c_{11}+c_{21} &= 0, \\ c_{12}+c_{22} &= 0, \end{aligned} \tag{3.4}$$

where a, b, c are estimates of α, β, γ respectively. Note that only three of the last four are independent, so that only five independent constraints in all are applied. With these constraints we can now solve the equations

$$m+a_i+b_j+c_{ij} = y_{ij}$$

to give

$$\begin{aligned} m &= y_{..}, \\ a_i &= y_{i.}-y_{..}, \\ b_j &= y_{.j}-y_{..}, \\ c_{ij} &= y_{ij}-y_{i.}-y_{.j}+y_{..}. \end{aligned} \tag{3.5}$$

Thus m is centred at the mean of all the estimated linear predictors, a_i is given by the deviation of a row mean from the grand mean, and

b_j similarly for a column. The interaction term c_{ij} represents the deviation of y_{ij}, the linear predictor for that cell, from one based on the additive components $m+a_i+b_j$.

The second set of constraints is that used in the program GLIM, and consists of

$$a_1 = b_1 = c_{11} = c_{12} = c_{21} = 0. \tag{3.6}$$

Here the top left-hand corner cell is taken as the 'baseline' for m, and in terms of the estimated linear predictors we have

$$\begin{aligned} m &= y_{11}, \\ a_2 &= y_{21}-y_{11}, \\ b_2 &= y_{12}-y_{11}, \\ c_{22} &= y_{22}-y_{12}-y_{21}+y_{11}. \end{aligned} \tag{3.7}$$

The remaining parameter estimates being zero by the constraints. Note that if $c_{ij} = 0$ then the as and bs given by the two formulae (3.5) and (3.7) become identical, and if further $a_i = b_j = 0$ then the ms become identical. This reflects the marginal relations between the terms.

It must be stressed again that constraints as exemplified by (3.4) and (3.6) are not a part of the model, but represent a convention whereby unique values for the estimates of intrinsically aliased parameters can be produced. Only the estimable quantities are relevant and those are independent of the constraint system imposed.

3.5.3 *Extrinsic aliasing*

The aliasing considered so far has been essentially a characteristic of the model rather than of the particular data set which is to give information about that model. However, aliasing can also occur because the particular covariate values in a data set contain linear dependencies. Suppose we have data for a 3×3 cross-classification in five out of the nine possible cells arranged as follows:

			B	
		1	2	3
	1	×	×	
A	2	×	×	
	3			×

Then writing a_i and b_j for the dummy variables for A and B we find that $a_3 = b_3$ in addition to $a_1+a_2+a_3 = b_1+b_2+b_3$. Thus, because of the particular configuration of the data, the two three-dimensional spaces A and B have a two-dimensional rather than one-dimensional subspace in common.

The aliasing reflects the fact that the five occupied cells can be split into two disconnected portions (of sizes 2×2 and 1×1) having no row or column in common. If we now move one of the occupied cells to produce the following configuration:

		B 1	2	3
A	1	×	×	
	2	×		
	3		×	×

the aliasing disappears along with the disconnectedness. This shows how extrinsic aliasing depends upon the pattern of the data, in contrast to intrinsic aliasing which is related to the formulation of the model.

3.5.4 *Functional relations in covariates*

Covariates may be functionally related without being linearly related. The simplest example is given by the covariates used in fitting a polynomial such as

$$\beta_0+\beta_1 x+\beta_2 x^2+\beta_3 x^3,$$

where the linear predictor contains terms in x, x^2 and x^3. The covariates x, x^2 and x^3 are linearly independent, so that no marginality relations hold; nonetheless there is often an implied ordering of the corresponding terms which has to be borne in mind in fitting models involving them.

Looking first at the terms β_0 and $\beta_1 x$ we need to ask when it makes sense not to use the sequence $\beta_0, \beta_0+\beta_1 x$ in fitting, but instead to begin with $\beta_1 x$ and no intercept. The answer is when $x = 0$ is a special point on the scale at which η must be zero. It is easy to give examples where this is not so; for example in an agricultural field trial with fertilizers, there will be some of the relevant nutrient already in the soil, so that zero fertilizer applied does not imply that none is available. Thus zero is *not* a special point.

Consider next the relationship between the terms $\beta_1 x$ and $\beta_2 x^2$. To fit terms β_0 and $\beta_2 x^2$ without $\beta_1 x$ implies that the maximum (or minimum) of the response occurs where $x = 0$. For if the x-scale might be measured just as well by $x+c$ as by x, then the response

$$\beta_0 + \beta_2 x^2$$

becomes

$$\beta_0 + \beta_2(x+c)^2 = (\beta_0 + \beta_2 c^2) + 2\beta_2 cx + \beta_2 x^2$$

and a linear term appears with $\beta_1 = 2\beta_2 c$. In general the position of the turning point in the response to x will *not* be a special point (namely zero) on the x-scale, so that the fitting of $\beta_2 x^2$ without $\beta_1 x$ is generally unhelpful.

A further example, involving more than one covariate, concerns the relation of a 'cross-term' $\beta_{12} x_1 x_2$ to the simple terms $\beta_1 x_1$ and $\beta_2 x_2$. To include the former without the latter is equivalent to assuming that the point (0, 0) is at a col (saddle point) in the response surface. Again there is usually no reason to postulate a special position for the origin, so that the linear terms must be included with the cross-term. It may, of course, happen that only the linear terms are required in the model, so that the quadratic ($\beta_2 x^2$) or the bilinear ($\beta_{12} x_1 x_2$) ones can be omitted, and these asymmetric relations between the first- and second-order terms can be thought of as constituting a functional marginality between terms. This functional marginality is not a true marginality, as defined in Section 3.5.1, because no linear dependences between covariates are involved, but it does reflect in the same way an ordering of the terms when models are being fitted.

3.6 Estimation

3.6.1 *The maximum-likelihood equations*

We shall deal primarily with maximum-likelihood estimation for all generalized linear models. For models with Normal errors and identity link, the log likelihood l is given by

$$-2l = N \ln(2\pi\sigma^2) + \sum_{i=1}^{n} (y_i - \mu_i)^2/\sigma^2,$$

where

$$\mu_i = \eta_i = \sum_{j=1}^{p} \beta_j x_{ij}.$$

Thus maximizing l is equivalent to choosing the β_j to minimize the sum of squares

$$\Sigma(y-\mu)^2.$$

Differentiating with respect to β_j and equating to zero, we find the estimating equations in the form

$$\sum_i (y_i - \hat{\mu}_i)x_{ij} = 0; \qquad j = 1, \ldots, p, \tag{3.8}$$

where

$$\hat{\mu}_i = \hat{\eta}_i = \sum_{j=1}^{p} b_j x_{ij},$$

the b_j being the ML estimates, $\hat{\eta}_i$ the fitted linear predictor, and $\hat{\mu}_i$ the fitted value for unit i. A useful way of regarding the equations (3.8) is to note that if a set of the xs in them are the dummy variates for a factor then $\Sigma_i y_i x_{ij}$ are just the sample marginal totals for that factor, and $\Sigma_i \hat{\mu}_i x_{ij}$ are similarly the fitted marginal totals. If an x is a continuous covariate we define the same quantities to be the corresponding margins for such a term. The maximum-likelihood equations can then be described as 'choose the bs to make the actual and fitted marginal totals the same for all terms in the model'.

3.6.2 *Geometrical interpretation*

Classical least squares has a geometrical interpretation which many find useful. We can regard the data $\mathbf{y}$ as a point in an N-dimensional sample space. For given values of β_j the fitted values $\boldsymbol{\mu}$ also represent a point in the same space. As the βs vary over all possible values they might take, the points $\boldsymbol{\mu}$ vary over a linear subspace or hyperplane called the *solution locus*. Any point on the solution locus corresponds to a perfect fit to the model, because if the data corresponded to such a point there is a set of β_j that will reproduce the N values exactly. Of course, in general, an actual set of data will not give a perfect fit so that the data point $\mathbf{y}$ does not lie in the solution locus. If $\boldsymbol{\mu}$ is a point on the solution locus then $\Sigma(y_i - \mu_i)^2$ is just the squared distance between $\mathbf{y}$ and $\boldsymbol{\mu}$. Thus the *maximum-likelihood* equations are equivalent to choosing values b_j for β_j to make $\hat{\boldsymbol{\mu}}$ in the solution locus as close as possible to $\mathbf{y}$.

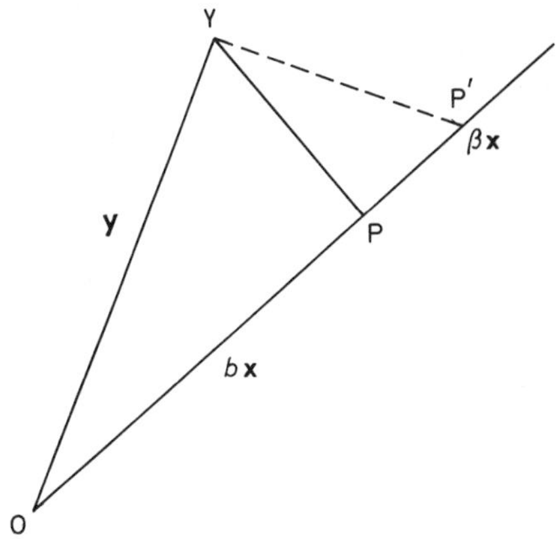

Fig. 3.3. *Least squares – the geometry for one parameter.*

For example, the model $\eta = \beta x$ represents a straight line through the origin (Fig. 3.3). The solution locus is $\beta\mathbf{x}$, i.e. points on the line through the origin in direction $\mathbf{x}$. The nearest point to $\mathbf{y}$ is found by dropping a perpendicular YP onto $\beta\mathbf{x}$. The coordinates of P are $b\mathbf{x}$, where b is the maximum-likelihood estimate of β, and the vector $\text{YP} = \mathbf{y} - b\mathbf{x}$ is the *residual vector*. The condition that OP and PY should be orthogonal is that

$$(\mathbf{y} - b\mathbf{x})^{\mathrm{T}}\mathbf{x} = 0,$$

i.e.

$$b = \frac{\mathbf{y}^{\mathrm{T}}\mathbf{x}}{\mathbf{x}^{\mathrm{T}}\mathbf{x}}.$$

The vector OP is the *projection* of OY on the direction $\mathbf{x}$.

3.6.3 *Information*

Further insight into the fit can be obtained by considering how the goodness of fit for a given $\boldsymbol{\mu}$, as measured by the size of $\Sigma(y-\mu)^2$, varies as we move along the solution locus. Let P′ be an arbitrary point on the solution locus, $\beta\mathbf{x}$ in above example. Then in the triangle YPP′ we have

$$(\mathbf{y} - \beta\mathbf{x})^{\mathrm{T}}(\mathbf{y} - \beta\mathbf{x}) = (\mathbf{y} - b\mathbf{x})^{\mathrm{T}}(\mathbf{y} - b\mathbf{x}) + (b - \beta)\mathbf{x}^{\mathrm{T}}\mathbf{x}(b - \beta),$$

expressing the Pythagorean relation. If we plot the l.h.s., which measures the discrepancy of the data from an arbitrary point on the solution locus, as a function of the parameter β, then we obtain a parabola with a minimum at $\beta = b$, the maximum-likelihood estimate, and a minimum discrepancy of $h = (\mathbf{y}-b\mathbf{x})^T(\mathbf{y}-b\mathbf{x})$ (Fig. 3.4).

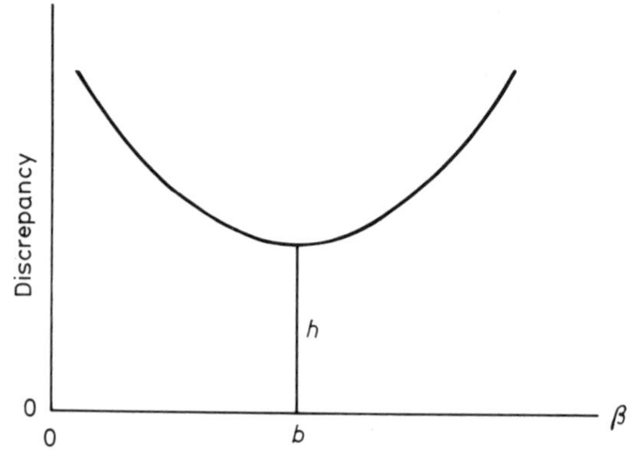

Fig. 3.4. *Information curve for one parameter.*

The second derivative at the minimum (and indeed elsewhere for a parabola) is given by $\mathbf{x}^T\mathbf{x}$. If we now restore the dispersion parameter σ^2 which divided the sum of squares then the second derivative becomes $\mathbf{x}^T\mathbf{x}/\sigma^2$, and this is Fisher's information on the parameter β. If the curvature is large the parabola is very steep-sided, so that small changes in β away from b produce large increases in the discrepancy, i.e. β is well determined.

The information is the ratio of two quantities. The numerator depends only on the model matrix, i.e. the values of the covariates in the model, and not at all on the dependent variate (y) values, while the denominator depends only on the variance of the error distribution of y. For this class of model the inverse of the information gives the theoretical sampling variance of the estimate b, i.e. $\text{var}(b) = \sigma^2/(\mathbf{x}^T\mathbf{x})$. Usually σ^2 requires to be estimated, either from replicates of the observations for the same x, or from the residual variance after fitting an adequate model.

3.6.4 *Model fitting with two covariates*

If there are two covariates, $\mathbf{x}_1$ and $\mathbf{x}_2$, say, then the solution locus is a plane defined by the points $\beta_1\mathbf{x}_1+\beta_2\mathbf{x}_2$ for varying β_1 and β_2.

The fitting of the model

$$\eta = \beta_1 x_1 + \beta_2 x_2$$

is represented geometrically by dropping a perpendicular from the data point $\mathbf{y}$ onto the $(\mathbf{x}_1, \mathbf{x}_2)$ plane. Fig. 3.5 shows the geometry connecting the individual fits of models $\beta_1 x_1$ and $\beta_2 x_2$ with the combined one.

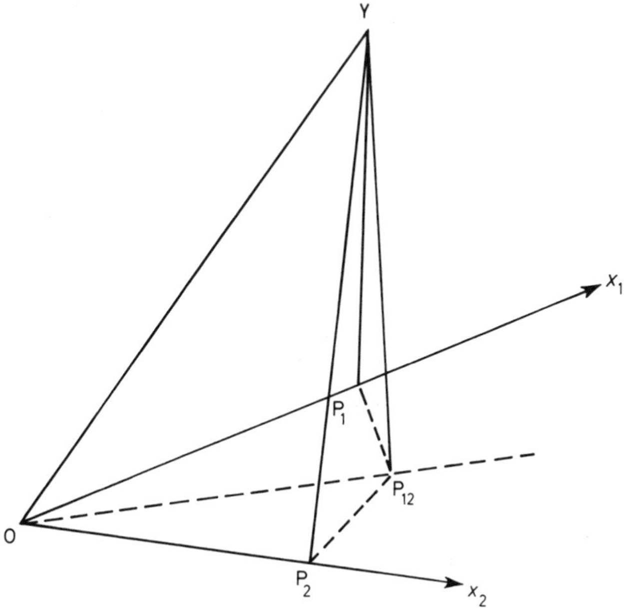

Fig. 3.5. *Least squares – the geometry for two parameters.*

The points P_1, P_2 and P_{12} are respectively the feet of the perpendiculars from $\mathbf{y}$ onto $\mathbf{x}_1$, $\mathbf{x}_2$ and the $(\mathbf{x}_1, \mathbf{x}_2)$ plane. The angles of OP_1P_{12} and OP_2P_{12} are both right angles (because OP_2Y, $OP_{12}Y$ and $P_2P_{12}Y$ are all right angles by definition); it follows that $OP_i^2 + P_iP_{12}^2 = OP_{12}^2$, $i = 1, 2$. We can thus express the projection of $\mathbf{y}$ on the $(\mathbf{x}_1, \mathbf{x}_2)$ plane in the two forms

$$OP_{12}^2 = OP_1^2 + P_1P_{12}^2 = OP_2^2 + P_2P_{12}^2.$$

Now

$$OP_1^2 = \text{sum of squares for } x_1 \text{ before } x_2,$$
$$OP_2^2 = \text{sum of squares for } x_2 \text{ before } x_1,$$
$$OP_{12}^2 = \text{sum of squares for } x_1 \text{ and } x_2,$$

so that

$$
\begin{aligned}
P_1P_{12}^2 &= \text{sum of squares for } x_2 \text{ after } x_1,\\
P_2P_{12}^2 &= \text{sum of squares for } x_1 \text{ after } x_2,\\
OY^2 &= \text{total sum of squares},\\
YP_{12}^2 &= \text{residual sum of squares}.
\end{aligned}
$$

The terms 'before' and 'after' are often replaced by 'ignoring' and 'eliminating'. Corresponding to the two sequences of fitting we have analyses of variance whose geometrical interpretation is

$$
\begin{aligned}
OY^2 &= OP_1^2 + P_1P_{12}^2 + P_{12}Y^2,\\
\text{total} &= (x_1 \text{ before } x_2) + (x_2 \text{ after } x_1) + \text{residual},
\end{aligned}
$$

and

$$
\begin{aligned}
OY^2 &= OP_2^2 + P_2P_{12}^2 + P_{12}Y^2,\\
\text{total} &= (x_2 \text{ before } x_1) + (x_1 \text{ after } x_2) + \text{residual}.
\end{aligned}
$$

In terms of the parameter estimates we have

$$
\begin{aligned}
OP_1 &= b_1\mathbf{x}_1,\\
OP_2 &= b_2\mathbf{x}_2,\\
OP_{12} &= b_{1.2}\mathbf{x}_1 + b_{2.1}\mathbf{x}_2,
\end{aligned}
$$

where b_1 and b_2 are the estimates for single-term models, and $b_{1.2}$ and $b_{2.1}$ the estimates for the joint model. There are several important cases of Fig. 3.5:

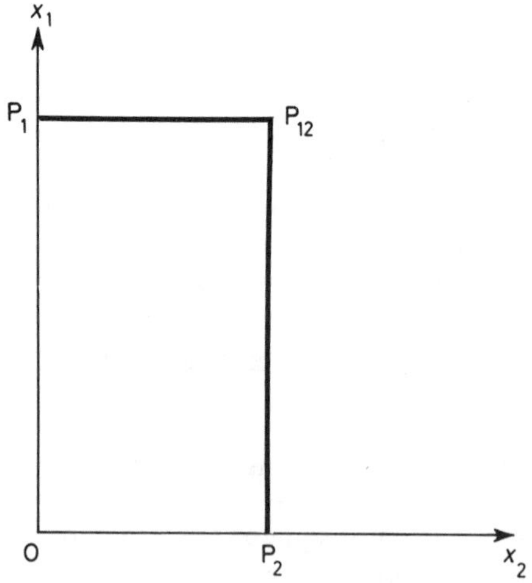

Fig. 3.6. *Least squares – projections for two orthogonal parameters.*

1. $\mathbf{y}$ is coplanar with $(\mathbf{x}_1, \mathbf{x}_2)$, so that Y and P_{12} coincide. The residual vector is null and the model gives a perfect fit.
2. $\mathbf{x}_1$ and $\mathbf{x}_2$ are orthogonal. The three feet of the perpendiculars P_1, P_2 and P_{12} form a rectangle with O so that $OP_1 = P_2P_{12}$ and $OP_2 = P_1P_{12}$ (Fig. 3.6). The order of fitting the terms in the joint model is irrelevant, and there is just one analysis of variance. Sums of squares and parameter estimates are unaffected by the order of fitting.
3. $\mathbf{y}$ is orthogonal to $\mathbf{x}_i$. The estimate $b_i = 0$ for a single-term model, but the corresponding joint estimate will not be zero unless $\mathbf{x}_1$ and $\mathbf{x}_2$ are orthogonal.

3.6.5 *The information surface*

If P is an arbitrary point $\beta_1\mathbf{x}_1 + \beta_2\mathbf{x}_2$ on the solution locus, then from the relation

$$YP^2 = YP_{12}^2 + P_{12}P^2$$

we obtain

$$\begin{aligned}(\mathbf{y}-\beta_1\mathbf{x}_1-\beta_2\mathbf{x}_2)^T(\mathbf{y}-\beta_1\mathbf{x}_1-\beta_2\mathbf{x}_2)\\ = (\mathbf{y}-b_1\mathbf{x}_1-b_2\mathbf{x}_2)^T(\mathbf{y}-b_1\mathbf{x}_1-b_2\mathbf{x}_2)\\ +[(b_1-\beta_1)^2\mathbf{x}_1^T\mathbf{x}_1+2(b_1-\beta_1)(b_2-\beta_2)\mathbf{x}_1^T\mathbf{x}_2\\ +(b_2-\beta_2)^2\mathbf{x}_2^T\mathbf{x}_2].\end{aligned}$$

(Note that b_1 and b_2 in this formula are from the *joint* fit, written above as $b_{1.2}$ and $b_{2.1}$ respectively.) The first term on the r.h.s. is the residual sum of squares and the second term (in square brackets) describes the *information surface*. The contours of this surface for different values of β_1 and β_2 are similar, similarly situated ellipses centred at (b_1, b_2) as shown in Fig. 3.7. The second-derivative matrix is given by

$$\begin{bmatrix}\mathbf{x}_1^T\mathbf{x}_1 & \mathbf{x}_1^T\mathbf{x}_2\\ \mathbf{x}_2^T\mathbf{x}_1 & \mathbf{x}_2^T\mathbf{x}_2\end{bmatrix}.$$

3.6.6 *Stability*

If θ, the angle between $\mathbf{x}_1$ and $\mathbf{x}_2$, is small, then P_{12} is unstable compared to P_1 and P_2, in the sense that while small perturbations of $\mathbf{x}_1$ and $\mathbf{x}_2$ may produce small individual perturbations of P_1 and P_2, together they may produce a large perturbation of P_{12}. This

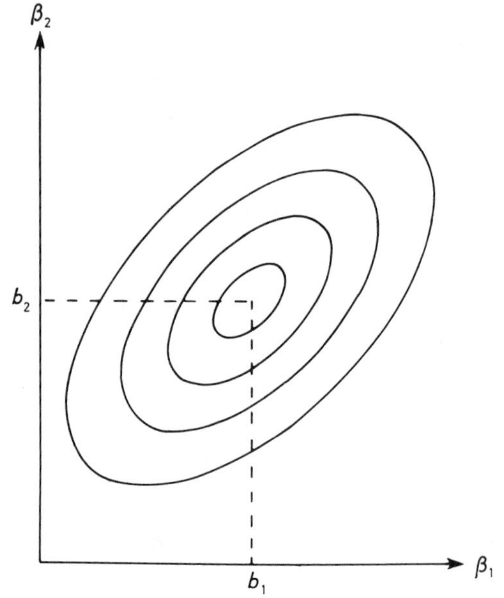

Fig. 3.7. *Least squares – information surface for two parameters.*

implies that the allocation of variation in Y due to regression of x_1 and x_2 is unstable. The information matrix is now nearly singular, because its determinant equals

$$\|\mathbf{x}_1\| \, \|\mathbf{x}_2\| \sin^2 \theta,$$

and the elliptical contours have one principal axis very long compared to the other. In consequence, relatively large changes in β_1 and β_2 along the direction of the long axis produce relatively small changes in the goodness of fit of the model.

3.7 Tables as data

It is common to find generalized linear models being fitted not to original data expressed in data-matrix form but to data that have already been summarized in the form of multi-way tables. In the process of tabulation the y-values for units having the same levels of the classifying factors are added together to form a table of totals; parallel tabulation of a vector of 1s gives the associated table of counts, showing how many units contribute to each cell total.

Division of totals by counts gives a table of means, and for continuous data it is usually this table that will be analysed, with the counts acting as prior weights. In surveys it is frequently the counts themselves that are of primary interest and methods of analysis of such tables are dealt with in Chapters 4 to 6.

In general the fitting of generalized linear models to data in the form of tables follows the same lines as with data in the original data-matrix form, but some topics deserve comment; none is in fact peculiar to tables, but discussion of them is usually in that context.

3.7.1 *Empty cells*

When a quantity is tabulated, be it a measurement or a count, an associated table of counts, as described above, is formed. This table is different in an important respect from a table formed by tabulating counts, namely the significance in it of zeros. If no unit in a data matrix contains a certain combination of levels of a set of classifying factors, then a table indexed by that set of factors will have no entry for the corresponding cell, but the table of associated counts will have a true zero for that cell. We shall use the term 'empty cell' for such a cell, implying that the set of units contributing to it is empty. There are two species of empty cell which it is important to distinguish; the first we call 'necessarily empty cell' and the second 'accidentally empty cell'. The term 'structural zero' is used by Bishop *et al.* (1975) for a necessarily empty cell in tables of counts, but the term is clearly unsatisfactory when applied to tables of measurements.

Necessarily empty cells occur when some combination of levels of the classifying factors is *a priori* impossible. Examples are the class of pregnant males or a self-fertilized cross from a self-sterile variety of plant. When all possible crosses are made between varieties of a self-sterile species, the diagonal cells, corresponding to the selfs, are all necessarily empty. If some varieties are cross-incompatible there may be off-diagonal necessarily empty cells as well. If a model is being fitted to a table of associated counts then necessarily empty cells must not be included as data. For other tables they cannot of course be included because there is no value for that cell; in general it makes no sense to calculate fitted values for such cells after a model has been fitted to the values from the non-empty cells.

An accidentally empty cell corresponds to a possible combination of the levels of the classifying factors, but it so happens that this combination has not occurred in any of the units contributing to the table. For associated tables of counts, the zero of an accidentally empty cell is informative about the relative frequency for that cell, and such zeros should be used in fitting models to the associated counts. For tabulated quantities there is again no value for such cells so that accidentally empty cells cannot be used in fitting. It will, however, generally make sense to give fitted values for such cells (provided that the model allows such values to be calculated).

It has been proposed (see, for example, Urquhart and Weeks, 1978) that standard models should not be fitted to tables that involve the population means of accidentally empty cells on the grounds that the data give us no information about such means. This would imply, for example, that an additive model for a two-way table of the form

$$A + B$$

should not be fitted if any cells are accidentally empty. The point has been discussed by Nelder (1982), who argues that such a rule is unnecessarily restrictive.

3.7.2 *Fused cells*

It sometimes happens that the position of a unit in a table is not known uniquely, though it is known to belong to one of a subset of cells. An example occurs in tables classified by genetic factors when several distinct genotypes produce the same phenotype, which is what is observed. What is available for analysis is then the total for the set of cells, fused into a single observable cell. Fused cells may also arise when the individual cells were observable, but for some reason the level of one or more factors was recorded with less precision than intended. The occurrence of fused cells results in an obvious loss of information and utilization of the data they contain may require prior knowledge of the relative frequency of occurrence in the unobserved component cells. Such knowledge is often available for genetic data.

3.8 Algorithms for least squares

For the classical linear models of this chapter the estimation process requires us to minimize

$$(\mathbf{y}-\mathbf{X}\boldsymbol{\beta})^{\mathrm{T}}(\mathbf{y}-\mathbf{X}\boldsymbol{\beta})$$

w.r.t. the parameters $\boldsymbol{\beta}$. Equating the derivatives to zero produces the *normal equations*

$$(\mathbf{X}^{\mathrm{T}}\mathbf{X})\hat{\boldsymbol{\beta}} = \mathbf{X}^{\mathrm{T}}\mathbf{y}. \tag{3.9}$$

If $\mathbf{X}$ is of less than full rank, because of aliasing among the parameters, then replacement of the inverse by a generalized inverse (Pringle and Rayner, 1971) yields estimates for all estimable contrasts among the βs. The (generalized) inverse also yields (apart from the dispersion parameter σ^2) the variances and covariances of the estimates.

There are two classes of method for deriving the estimates $\hat{\boldsymbol{\beta}}$. One involves the explicit formation of $\mathbf{X}^{\mathrm{T}}\mathbf{X}$ and its inversion, while the other involves the decomposition of the model matrix $\mathbf{X}$. For both classes it is usual to express both the dependent variable and the columns of the $\mathbf{X}$ matrix about their means. This is equivalent to including the intercept (or grand mean) in the model being fitted. Within each class there are further subdivisions, which we now briefly outline.

3.8.1 *Methods based on the information matrix*

There are two main methods which utilize the information matrix $\mathbf{X}^{\mathrm{T}}\mathbf{X}$: Gaussian elimination and Choleski decomposition.

A modern form of Gaussian elimination, due to Beaton (1964), uses a symmetric sweep operator. This operator, when applied to the kth row and column of a positive-definite symmetric matrix $\mathbf{A}$, will be denoted by $\mathscr{S}_k$, and is defined by the relations

$$a_{ij} \to a_{ij} - \frac{a_{ik}a_{jk}}{a_{kk}}; \qquad i \neq k, j \neq k,$$

$$a_{ik} \to \frac{a_{ik}}{|a_{kk}|}; \qquad i \neq k,$$

$$a_{kk} \to -\frac{1}{a_{kk}}.$$

It is easily shown that $\mathscr{S}_k \mathscr{S}_k \mathbf{A} = \mathbf{A}$, i.e. that a second application of the symmetric sweep restores the original form. The statistical interpretation of the symmetric sweep is as follows. Let $\mathbf{A} = \mathbf{X}^{\mathrm{T}}\mathbf{X}$ be a $p \times p$ matrix of sums of squares and products of the variates $\mathbf{x}_1, \ldots, \mathbf{x}_p$, and suppose the sweeps $\mathscr{S}_1, \mathscr{S}_2, \ldots, \mathscr{S}_k$ have been applied to the first k ($k < p$) rows/columns; $\mathbf{A}$ has now been reduced to the form shown in Fig. 3.8. The component $\mathbf{R}$ now contains the residual sum-of-squares matrix for the unswept variates $\mathbf{x}_{k+1}, \ldots, \mathbf{x}_p$ after regressing them on $\mathbf{x}_1, \ldots, \mathbf{x}_k$; the rows of $\mathbf{B}$ are the regression coefficients of those regression equations, while $-\mathbf{V}$ is minus the common unscaled covariance matrix for the regressions.

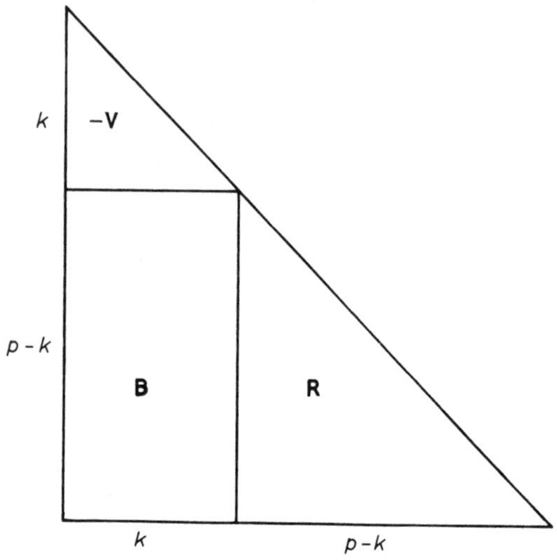

Fig. 3.8. *The matrix of sums of squares and products after the symmetric sweep has been applied to the first k rows/columns.*

If the original $\mathbf{X}^{\mathrm{T}}\mathbf{X}$ is exactly or nearly singular, this will usually show up during the sweeping process by the appearance of a pivot that is small compared to its original value (see Clarke, 1982). Statistically this means that there is (almost) no information about the corresponding parameter, given that the previously fitted parameters are in the model. Without any information about a parameter we can only omit it from the model; algebraically this is equivalent to setting its value to zero with variance zero. Algorithmically, we do not sweep the row/column but mark it to show its special status.

If the parameter is subsequently to be removed from the model then again no sweep is done; if its pivot becomes substantial at any point the parameter can then be included in the model.

The Choleski decomposition finds a lower-triangular $p \times p$ matrix $\mathbf{L}$ which satisfies

$$\mathbf{X}^{\mathrm{T}}\mathbf{X} = \mathbf{L}\mathbf{L}^{\mathrm{T}}.$$

$\mathbf{L}$ is thus a form of matrix square root of the SSP (sum of squares and products) matrix $\mathbf{X}^{\mathrm{T}}\mathbf{X}$. The inversion of $\mathbf{X}^{\mathrm{T}}\mathbf{X}$ is achieved via the formula

$$(\mathbf{X}^{\mathrm{T}}\mathbf{X})^{-1} = (\mathbf{L}^{-1})^{\mathrm{T}}\mathbf{L}^{-1}.$$

There is a simple inversion algorithm for triangular matrices, and the inversion can be combined with subsequent multiplication by the transpose; see, for example, Healy (1968) and Chambers (1977) for the details of the algebra. Again generalized inverses can be obtained for restricted models by setting any row with a small pivot to zero.

From the numerical-analytic point of view, methods which utilize the $\mathbf{X}^{\mathrm{T}}\mathbf{X}$ matrix suffer from the disadvantage that if the condition number of $\mathbf{X}$ (which measures its approach to singularity) is large, then the condition number of $\mathbf{X}^{\mathrm{T}}\mathbf{X}$, being the square of that of $\mathbf{X}$, increases the instability resulting from rounding errors in the calculations. The second class of methods is designed to avoid the formation of $\mathbf{X}^{\mathrm{T}}\mathbf{X}$ altogether.

3.8.2 *Direct-decomposition methods*

These methods decompose the $n \times p$ model matrix $\mathbf{X}$ into the product of an $n \times n$ orthogonal matrix $\mathbf{Q}$ and an $n \times p$ matrix $\mathbf{R}$ of the form

$$\mathbf{R} = \begin{bmatrix} \mathbf{R}_1 \\ \mathbf{0} \end{bmatrix},$$

where $\mathbf{R}_1$ is $p \times p$ upper-triangular. We can derive a statistical interpretation of this decomposition as follows: If $\mathbf{y}$ is our vector of observations with mean $\mathbf{X}\boldsymbol{\beta}$ and variance $\sigma^2\mathbf{I}$, consider its replacement by a set of contrasts $\mathbf{u}$ defined by $\mathbf{u} = \mathbf{Q}^{\mathrm{T}}\mathbf{y}$ where $\mathbf{Q}$ is an orthogonal matrix. Then

$$\begin{aligned} \mathrm{E(U)} = \mathbf{Q}^{\mathrm{T}}\mathrm{E(Y)} &= \mathbf{Q}^{\mathrm{T}}\mathbf{X}\boldsymbol{\beta} = \mathbf{Q}^{\mathrm{T}}\mathbf{Q}\mathbf{R}\boldsymbol{\beta} \\ &= \mathbf{R}\boldsymbol{\beta} \quad = \begin{bmatrix} \mathbf{R}_1\boldsymbol{\beta} \\ \mathbf{0} \end{bmatrix} \end{aligned}$$

and

$$\text{var}(\mathbf{U}) = \mathbf{Q}^T\mathbf{I}\mathbf{Q}\sigma^2 = \mathbf{I}\sigma^2.$$

Thus the last $(n-p)$ of the u_i have expectation zero, and so give no information about $\boldsymbol{\beta}$. The first p of the u_i form the vector $\mathbf{u}_1$ whose length equals the number of parameters, and hence the least-squares solution comes from solving

$$\mathbf{u}_1 = \mathbf{R}_1\mathbf{b},$$

which is easy because $\mathbf{R}_1$ is upper-triangular. Note that we do not need to know $\mathbf{Q}$ explicitly because

$$\mathbf{R}^T\mathbf{u} = \mathbf{R}^T\mathbf{Q}^T\mathbf{y} = \mathbf{X}^T\mathbf{y}.$$

Hence $\mathbf{u}$ can be obtained directly from solving this equation. Note also that

$$\mathbf{R}^T\mathbf{R} = \mathbf{R}^T\mathbf{Q}^T\mathbf{Q}\mathbf{R} = \mathbf{X}^T\mathbf{X},$$

so that $\mathbf{R}$ is the upper-triangular Choleski square-root matrix and so equal to the transpose of $\mathbf{L}$.

There are three methods extant for finding $\mathbf{Q}$ and $\mathbf{R}$; these are associated with the names of

Householder	($\mathbf{Q}$ is a product of reflections),
Givens	($\mathbf{Q}$ is a product of rotations),
Gram–Schmidt	(successive orthogonalization).

A Householder reflection takes the matrix form

$$\mathbf{I} - 2\mathbf{u}\mathbf{u}^T,$$

where $\mathbf{u}$ is an n-vector of unit length ($\mathbf{u}^T\mathbf{u} = 1$). It is possible, given a general vector $\mathbf{x}$, to choose $\mathbf{u}$ so that after reflection all $\mathbf{x}$ elements except the first are transformed to zero. In the Householder decomposition $\mathbf{u}_1$ is chosen to reduce the first column of $\mathbf{X}$ to this form. A second vector $\mathbf{u}_2$ is chosen to reduce elements 3 to n of the second column to zero; since these elements in the first column are already all zero, this reflection leaves them as zero. The process continues on elements $i+1$ to n of the ith column for $i = 3, \ldots, n-1$. If $\mathbf{Q}_i = \mathbf{I} - 2\mathbf{u}_i\mathbf{u}_i^T$ then the matrix $\mathbf{Q}$ is defined by

$$\mathbf{Q} = \mathbf{Q}_{n-1}\mathbf{Q}_{n-2}\cdots\mathbf{Q}_2\mathbf{Q}_1,$$

the product of the successive reflections.

Givens rotations are planar rotations through an angle θ. A single rotation is applied to two elements in a vector and θ chosen to transform one of those elements to zero. If we call G_{ijk} the rotation that replaces the ith and jth rows of $\mathbf{X}$ by linear combinations which make zero the kth element in row j, then the sequence

$$((G_{kjk}, j = k+1 \text{ to } n),\ k = 1 \text{ to } p)$$

annihilates the elements by columns, like Householder, but setting one element to zero at a time. The sequence

$$((G_{kjk}, k = 1 \text{ to } \min(j-1, p)),\ j = 2 \text{ to } n)$$

annihilates by rows, and can be useful if the $\mathbf{X}$ matrix is more easily processed by rows than by columns. The idea of rotations goes back to Jacobi, but the Givens sequence of rotations ensures that previously formed zeros stay zero.

The Gram–Schmidt method relies on successive orthogonalization of the columns of $\mathbf{X}$. The preferred algorithmic form is due to Björck (1967) and begins by forming

$$\mathbf{q}_1 = \frac{\mathbf{x}_1}{\|\mathbf{x}_1\|}$$

and

$$\mathbf{q}_j = \mathbf{x}_j - (\mathbf{q}_1^{\mathrm{T}}\mathbf{x}_j)\,\mathbf{q}_1; \qquad j = 2 \text{ to } p,$$

then

$$r_{1i} = \mathbf{q}_1^{\mathrm{T}}\mathbf{x}_i.$$

This process is repeated, using in the second stage the vectors $\mathbf{q}_2, \ldots, \mathbf{q}_p$, etc.

Statistically we regress columns 2 to p on column 1 and replace them by the vectors of residuals, repeating with columns 3 to p on column 2, etc. $\mathbf{Q}$ is here an $n \times p$ orthonormal matrix and $\mathbf{R}$ a $p \times p$ upper-triangular matrix; the first j columns of $\mathbf{Q}$ span the same space as the first j columns of $\mathbf{X}$ for $j = 1, \ldots, p$. Regression of $\mathbf{y}$ on the orthogonalized variables is easy because of the orthogonality.

Updating models with the direct-decomposition methods is somewhat less convenient than using the symmetric sweep; for example if a column of $\mathbf{X}$ is deleted, all the columns of $\mathbf{Q}$ and $\mathbf{R}$ to the right of it have to be recalculated. Details of updating with the Givens algorithm are given in Clarke (1981).

3.8.3 *Extension to generalized linear models*

In Chapter 2 we showed how to reduce the estimation problem for generalized linear models to iterative weighted least squares. To adapt the algorithms discussed above we must (a) introduce weights and adjusted dependent variates and (b) allow iteration. The introduction of weights is straightforward; in the algorithms using the information matrix we replace $\mathbf{X}^T\mathbf{X}$ by $\mathbf{X}^T\mathbf{W}\mathbf{X}$, where $\mathbf{W}$ is the diagonal matrix of weights, while in the QR algorithms we replace $\mathbf{X}$ by $\mathbf{W}^{\frac{1}{2}}\mathbf{X}$. The only new feature in the behaviour of the algorithms is a pseudo-aliasing that can arise when the changing weights produce so little information on a parameter that the diagonal element falls below the tolerance set in the algorithm. However, removal of this parameter will now be found to increase the deviance sharply, showing a difference in behaviour from true aliasing. The fit just before the parameter becomes apparently aliased will often be adequate.

Iteration requires the definition of convergence of the process. If all parameter estimates are bounded, a simple monitoring of the progress of the deviance is sufficient. Convergence is usually rapid, though divergence may occur, particularly with ill-fitting models using non-canonical links. If some parameter estimates are unbounded, convergence, as measured by the deviance, may be very slow. It is usually best to halt the iteration after, say, 10 cycles and inspect the parameter estimates, from which it will be clear which are tending to $\pm\infty$. Action can then be taken to omit a subset of the data or modify the model, or both.

3.9 Selection of covariates

We discussed in Chapter 1 the justification for seeking a parsimonious model to represent a set of data. Parsimony implies, among other things, selection from among the many possible forms of linear predictor; in a survey on, for example, the incidence of a disease, large numbers of covariates may be available, describing perhaps age structure of the populations involved, their dietary and smoking habits, aspects of their environment, etc. The selection of a useful set of covariates to form the basis of a parsimonious model then becomes a non-trivial exercise. There are both statistical and computing problems, the latter arising from the 'combinatorial explosion' when all possible combinations of subsets of covariates are to be tested for inclusion in the model.

The statistical problem is that of defining the balance between two effects of including a new term in the linear predictor. The good effect may be a reduction in the discrepancy between the data and the fitted values in the model; the bad effect is a less parsimonious model. At the two extremes there is no problem. On the one hand if the addition of a single parameter reduces the residual mean square to, say, a third of its original value we have no hesitation in including it, particularly if the residual number of degrees of freedom is large. On the other hand if such an addition causes no reduction, parsimony wins and we exclude it. It is the intermediate cases that cause the difficulty. If we imagine the presence of many covariates which in fact are all irrelevant, then statistical accidents will produce a few false positives which appear to influence the response; such considerations underlie criteria for acceptance of a term based on probability levels for, say, the corresponding F-statistic. Another approach is based on the idea of providing the best prediction of response values over a set of covariate values, and yet another uses a criterion based on a measure of information. Atkinson (1981b) points out that all these procedures can be represented (in our notation) as special cases of minimizing the expression

$$Q = D + \alpha q \phi, \tag{3.10}$$

where D is the deviance for a model, q is the number of estimable parameters in the linear predictor, ϕ is the dispersion parameter and α is either a constant or a function of N, the number of units of data. The first term in Q measures the discrepancy (the smaller the better) while the second increases with the number of parameters (the more the worse). Use of Q presumes a knowledge of ϕ, the dispersion parameter. For generalized linear models with Poisson or binomial errors $\phi = 1$ (assuming no over-dispersion), but otherwise ϕ will not in general be known. When comparing a series of models we have the option of replacing ϕ either by a single estimate $\hat{\phi}$ for all the models (obtaining it perhaps from the most complex one) or by separate $\hat{\phi}_i$ for each model; the former seems preferable in practice.

If two models differ by one parameter then the use of the 5% point of the corresponding t- or F-distribution to assess whether the extra parameter should be included is equivalent to setting $\alpha \simeq 4$ (assuming adequate d.f. for the estimate of ϕ). Criteria based on prediction errors (Akaike, 1969; Mallows, 1973) lead to $\alpha = 2$, while for Normal models an argument based on maximum posterior probabilities leads to $\alpha(N) = \ln N$. Atkinson (1981b) suggests that the range $\alpha = 2$ to 6 may provide 'a set of plausible initial models for further analysis'.

The computing problem may be specified as 'find the best s subsets of the covariates of size r'. If k, the total number of covariates, is reasonably small, say $k \leqslant 12$, the best subsets for each r from 1 to $(k-1)$ can be found by complete enumeration. For somewhat larger k, say up to 35, tree-search methods, using short-cuts, are feasible (Furnival and Wilson, 1974). Approximate methods for generating a single 'optimum' subset include: (i) *forward selection*, whereby at each stage the best unselected covariate satisfying the criterion is included next until none remain; (ii) *backward elimination*, which begins with the full set and eliminates the worst covariates one by one until all remaining are necessary; and (iii) *stepwise regression* (Efroymson, 1960), which combines the two previous procedures, following backward elimination by forward selection until both fail.

Unthinking use of automatic procedures has frequently been criticized; clearly the notion that a particular subset is optimum is hard to sustain when many other subsets give almost as good fits. It may also happen that some covariates are much more expensive to measure than others, something that is not reflected in a criterion based on statistical considerations. Again the published algorithms for complete enumeration do not take account of marginality constraints between qualitative terms or functional marginality between polynomial terms. Furthermore, certain factors (e.g. treatment effects and block effects) would normally be kept in the model whether statistically significant or not.

Further modification of the procedures is required for models requiring iterative solution, because of the presence of weights and adjusted dependent variates which are both functions of the fitted values. An approximate procedure, which appears to work well in practice, involves doing the full iterative fit for a large but well fitting model, afterwards following the same algorithms as for the non-iterative case, i.e. keeping the weights and adjusted dependent variate fixed. The iterative fit may be recalculated at intervals as a check.

3.10 Bibliographic notes

For a history of least squares, see a series of papers by Harter summarized in Harter (1976). Among the many texts on linear models and regression analysis see, for example, Draper and Smith (1981), Mosteller and Tukey (1977), Plackett (1960), Searle (1971), Seber (1977), Sprent (1969) and Williams (1959). Model formulae for linear predictors were introduced by Nelder (1965a, b) and developed by Wilkinson and Rogers (1973). Aliasing and marginality are discussed by Nelder (1977). For numerical methods for least squares see Lawson and Hanson (1974), Gentleman (1974a, b), Chambers (1977) and Wampler (1979). The statistical problems of covariate selection are discussed by Akaike (1973), Mallows (1973), Stone (1977), and summarized by Atkinson (1981b). For the computing aspect see Efroymson (1960) (the original 'stepwise regression' paper), Beale (1970), Stewart (1973) and Furnival and Wilson (1974).

CHAPTER 4

Binary data

4.1 Introduction

4.1.1 *Binary responses*

Suppose that for each individual or experimental unit the response, Y, can take only one of two possible values, denoted for convenience by 0 and 1 respectively. Observations of this nature arise, for example, in medical trials where, at the end of the trial period, the patient has either recovered ($Y = 1$) or has not ($Y = 0$). Clearly we could have intermediate values associated with different degrees of recovery (see Chapter 5), but for the moment this possibility will be ignored. We may write

$$\text{pr}(Y_i = 0) = 1 - \Pi_i; \qquad \text{pr}(Y_i = 1) = \Pi_i. \tag{4.1}$$

In most investigations, whether they be designed experiments or observational studies, we have, associated with each individual or experimental unit, a vector of covariates or explanatory variables $(x_1, \ldots, x_p)$. In a designed experiment this vector of covariates generally consists of indicator variables associated with blocking and treatment factors, together with quantitative covariates concerning various aspects of the experimental material. In non-randomized studies the vector of covariates would consist of measured variables thought likely to influence the probability of a positive response. The principal objective is to investigate the relationship between the response probability Π and the explanatory variables $x_1, \ldots, x_p$. Often, of course, a subset of the xs is of primary importance but due allowance must also be made for the effect of the remaining covariates.

4.1.2 *Grouped data*

Suppose that for the ith combination of experimental conditions $(x_{ii}, \ldots, x_{ip}), i = 1, \ldots, N$, observations are available on n_i individuals. We now say that the observations are grouped, the group sizes being $n_1, \ldots, n_N$. The observations thus have the form $y_1/n_1, y_2/n_2, \ldots, y_N/n_N$ where y_i is the number of positive responses out of n_i trials. Here the response is restricted to lie in the range $0 \leqslant y_i \leqslant n_i$ and the vector $\mathbf{n} = (n_1, \ldots, n_N)$ will be called the binomial denominator. Ungrouped data comprise a special case with $n_1 = \ldots = n_N = 1$.

The distinction between grouped and ungrouped data is important for two reasons:

1. Some methods of analysis appropriate to grouped data are not applicable to ungrouped data.
2. Asymptotic approximations for grouped data can be based on one of two distinct asymptotes, either $\mathbf{n} \to \infty$ or $N \to \infty$. The latter asymptote only is appropriate for ungrouped data.

If the observations on all individuals are independent and if the probability of a positive response is constant for all individuals in the same group then the distribution of Y_i given n_i is binomial with index n_i and parameter Π_i. The likelihood function under these conditions is particularly simple; see, for example, Section 4.2.1. In many applications, however, observations in different groups may be independent or at least uncorrelated but there is reason to expect positive correlation within groups. Under these circumstances the variance of Y_i will be larger than that under independence, namely $n_i \Pi_i(1-\Pi_i)$: it is convenient then to make the assumption that

$$\operatorname{var}(Y_i) = \sigma^2 n_i \Pi_i(1-\Pi_i); \qquad i = 1, \ldots, N, \tag{4.2}$$

with $\sigma^2 > 1$, and to estimate σ^2 from the data. This phenomenon, known as over-dispersion, arises frequently in applications. In principle, of course, the data could be under-dispersed, i.e. we could have $\sigma^2 < 1$, but under-dispersion seems less common in applications. We deal with under-dispersion and over-dispersion by the use of quasi-likelihood functions in Section 4.2.2.

4.1.3 *Properties of the binomial distribution*

Although the assumption of binomial variation is not required to justify most of the techniques used here, the binomial model does play an important role in the analysis of binary response data. For completeness, therefore, we describe some of the more important properties of this distribution. It is convenient to use the notation $B(n, \Pi)$ for the binomial distribution with index n and probability Π and $N(\mu, \sigma^2)$ and $P(\mu)$ for the Normal and Poisson distributions respectively. Proofs of many of the statements made below can be found in standard texts, e.g. Cox (1970) and Plackett (1981).

Suppose $Y \sim B(n, \Pi)$. The first two moments of Y are

$$E(Y) = n\Pi \qquad \text{and} \qquad \operatorname{var}(Y) = n\Pi(1-\Pi).$$

All cumulants are $O(n)$: the third and fourth are

$$\kappa_3 = n\Pi(1-\Pi)(1-2\Pi)$$

$$\text{and} \qquad \kappa_4 = n\Pi(1-\Pi)[(1-6\Pi(1-\Pi)].$$

It follows from the fact that all cumulants are $O(n)$ that

$$(Y-n\Pi)/\sqrt{[n\Pi(1-\Pi)]} \sim N(0, 1)+O_p(n^{-\frac{1}{2}}).$$

The rate of convergence to Normality is governed primarily by the standardized third cumulant and is therefore fastest if $\Pi = 0.5$, corresponding to symmetry. Provided that a correction is made for continuity, the error of the Normal approximation is $O(n^{-1})$ in the symmetric case and $O(n^{-\frac{1}{2}})$ otherwise.

The Poisson approximation is obtained in the limit as $n \to \infty$ with $n\Pi = \mu$ held fixed. The error involved in using the Poisson approximation to the binomial appears to be $O(n^{-1})$.

For large n the approximate variance-stabilizing transform for Y is

$$Z = \arcsin[(Y/n)^{\frac{1}{2}}],$$

which is known as the angular transform. The approximate mean and variance are

$$E(Z) \simeq \arcsin(\Pi^{\frac{1}{2}}) \qquad \text{and} \qquad \operatorname{var}(Z) \simeq (4n)^{-1}.$$

The variance-stabilizing transform is distinct from the transformation to symmetry (Cox and Snell, 1968) which involves the incomplete beta function. An alternative transform to symmetry can be based on the deviance function (see Section 2.3). By a tedious argument involving a Taylor series expansion it can be shown that the moments of the statistic

$$\pm\{2Y\ln(Y/\mu)+2(n-Y)\ln[(n-Y)/(n-\mu)]\}^{\frac{1}{2}} + \{(1-2\Pi)/[n\Pi(1-\Pi)]^{\frac{1}{2}}\}/6$$

differ from those of N(0, 1) by $O(n^{-1})$. The sign used is that of $Y-\mu$ and the transform is a monotonic function of $Y-\mu$. This latter transform induces symmetry and stabilizes the variance simultaneously. It arises in a fairly natural way in fitting algorithms and can be used as a residual when μ is replaced by $\hat{\mu}$.

The empirical logistic transform of Y is given by

$$Z = \ln[(Y+\tfrac{1}{2})/(n-Y+\tfrac{1}{2})]$$

(Cox, 1970, p. 33). Here we have added $\frac{1}{2}$ to both numerator and denominator to reduce bias in the sense that if we set $\lambda = \ln[\Pi/(1-\Pi)]$ then

$$E(Z) = \lambda + O(n^{-2}).$$

Gart and Zweifel's (1967) results support the estimation of var(Z) by $(y+\frac{1}{2})^{-1}+(n-y+\frac{1}{2})^{-1}$. Because of the nature of the asymptotic argument the transform is useful only if all the binomial indices are large.

4.1.4 *Models for binary responses*

To investigate the relation between the response probability Π and the covariate vector $(x_1, \ldots, x_p)$ it is convenient, though perhaps not absolutely necessary, to construct a formal model thought capable of describing, at least approximately, the relation between changes in $(x_1, \ldots, x_p)$ and changes in Π. This formal model will usually embody assumptions such as zero correlation, lack of interaction, and so on. In applications these assumptions cannot be taken for granted and should, if possible, be checked. Furthermore the behaviour of the model should, as far as possible, be consistent with known physical, biological or mathematical laws.

Linear models play an important role in both applied and theoretical work. We suppose therefore that the dependence of Π on $(x_1, \ldots, x_p)$ is through the linear combination

$$\eta = \sum_{j=1}^{p} x_j \beta_j \qquad (4.3)$$

for some unknown coefficients $\beta_1, \ldots, \beta_p$. Unless restrictions are imposed on $\beta_1, \ldots, \beta_p$ we have $-\infty < \eta < \infty$, so that to express Π as the linear combination (4.3) would be inconsistent with the laws of probability. A simple way of avoiding this problem is to use a transformation $g(\Pi)$ which maps the unit interval (0, 1) onto the real line $(-\infty, \infty)$. This leads to instances of generalized linear models with

$$g(\Pi_i) = \eta_i = \sum_{j=1}^{p} x_{ij}\beta_j; \qquad i = 1, \ldots, N. \tag{4.4}$$

A wide choice of link functions $g(\Pi)$ is available. Three functions commonly used in practice are:

1. the logit function

$$g_1(\Pi) = \ln[\Pi/(1-\Pi)],$$

2. the probit function

$$g_2(\Pi) = \Phi^{-1}(\Pi),$$

3. the complementary log–log function

$$g_3(\Pi) = \ln[-\ln(1-\Pi)].$$

Here $\Phi^{-1}(\Pi)$ is the inverse of the standard Normal integral and the logit function is the inverse of a similar symmetric cumulative distribution function. The complementary log–log function is the inverse of the cumulative distribution of the extreme-value distribution. Unlike $g_1(\Pi)$ and $g_2(\Pi)$ it is not symmetric about $\Pi = 0.5$, so that

$$g_3(\Pi) \neq -g_3(1-\Pi).$$

Table 4.1 compares the three functions standardized for convenience at the 20% and 80% points. The logit and probit functions are symmetric and, appropriately standardized, are very similar over the range $0.1 \leqslant \Pi \leqslant 0.9$. For this reason it is usually difficult to discriminate between them on the grounds of goodness of fit; see, for example, Chambers and Cox (1967). For small values of Π the complementary log–log function is close to the logistic (a fact obscured by the standardization imposed in constructing Table 4.1), but as Π approaches 1, it tends much more slowly to infinity than either the logit or probit transforms. We shall be concerned mostly with the logit function because of its simple interpretation as the logarithm of the odds ratio. The asymptotic methods of Section 4.3

Table 4.1 *A comparison of three link functions*

Π	*Logit* $0.6071g_1(\Pi)$	*Probit* $g_2(\Pi)$	*Complementary l*… $0.4362+0.8519g$…
0.001	−4.193	−3.090	−5.488
0.01	−2.790	−2.330	−3.483
0.05	−1.788	−1.645	−2.094
0.10	−1.334	−1.285	−1.481
0.20	−0.842	−0.842	−0.842
0.50	0.000	0.000	0.124
0.80	0.842	0.842	0.842
0.90	1.334	1.285	1.147
0.95	1.788	1.645	1.371
0.99	2.790	2.330	1.737
0.999	4.193	3.090	2.083

apply quite generally whatever the link function used. A further advantage, though a theoretical one, of the logit linear model is that it often presents a simple exact analysis by conditioning.

4.1.5 *Retrospective studies*

One important theoretical advantage of the logistic function over all other possible transforms is that differences on the logistic scale can be estimated regardless of whether the data are sampled prospectively or retrospectively. Suppose for example that a population is classified according to two binary variables $(D, \bar{D})$ referring to the presence or absence of disease and $(X, \bar{X})$ referring to exposure or non-exposure. Let the corresponding proportions in the population be given as follows:

		Disease status		
		$\bar{D}$	D	*Total*
Exposure status	$\bar{X}$	Π_{00}	Π_{01}	$\Pi_{0.}$
	X	Π_{10}	Π_{11}	$\Pi_{1.}$
	Total	$\Pi_{.0}$	$\Pi_{.1}$	1

In a prospective study we fix the totals associated with exposure. The corresponding logits are $\ln(\Pi_{01}/\Pi_{00})$ for the unexposed group and

$\ln(\Pi_{11}/\Pi_{10})$ for the exposed group. The difference

$$\Delta = \ln(\Pi_{11}\Pi_{00}/\Pi_{01}\Pi_{10})$$

could also be estimated by sampling retrospectively from the two disease groups $\bar{D}$ and D because

$$\Delta = \ln(\Pi_{11}/\Pi_{01}) - \ln(\Pi_{10}/\Pi_{00}).$$

In other words, even if the data are sampled retrospectively, estimates of the effect of treatment on the logistic scale may be found by application of the logistic model to $\mathrm{pr}(D|X)$.

More generally, following Armitage (1971) and Breslow and Day (1980), we may write

$$\mathrm{pr}(D|\mathbf{x}) = \exp(\alpha + \boldsymbol{\beta}^{\mathrm{T}}\mathbf{x})/[1 + \exp(\alpha + \boldsymbol{\beta}^{\mathrm{T}}\mathbf{x})]$$

for the probability of disease given an arbitrary vector of covariates $\mathbf{x}$. Suppose now that the data are sampled retrospectively. Introduce the dummy variable Z corresponding to whether or not an individual is sampled and denote the sampling proportions by

$$\Pi_0 = \mathrm{pr}(Z = 1|D) \qquad \text{and} \qquad \Pi_1 = \mathrm{pr}(Z = 1|\bar{D}).$$

It is assumed here that the sampling proportions depend only on D and not on $\mathbf{x}$. Now use Bayes's theorem to compute

$$\begin{aligned}\mathrm{pr}(D|Z = 1, \mathbf{x}) &= \frac{\mathrm{pr}(Z = 1|D, \mathbf{x})\,\mathrm{pr}(D|\mathbf{x})}{\mathrm{pr}(Z = 1|D, \mathbf{x})\,\mathrm{pr}(D|\mathbf{x}) + \mathrm{pr}(Z = 1|\bar{D}, \mathbf{x})\,\mathrm{pr}(\bar{D}|\mathbf{x})} \\ &= \frac{\Pi_1 \exp(\alpha + \boldsymbol{\beta}^{\mathrm{T}}\mathbf{x})}{\Pi_0 + \Pi_1 \exp(\alpha + \boldsymbol{\beta}^{\mathrm{T}}\mathbf{x})} \\ &= \frac{\exp(\alpha^* + \boldsymbol{\beta}^{\mathrm{T}}\mathbf{x})}{1 + \exp(\alpha^* + \boldsymbol{\beta}^{\mathrm{T}}\mathbf{x})},\end{aligned}$$

where $\alpha^* = \alpha + \ln(\Pi_1/\Pi_0)$. In other words, although the data have been sampled retrospectively, the logistic model continues to apply with exactly the same βs but with a different intercept. Thus the methods described here in the context of prospective studies can be applied to retrospective studies provided that the intercept is treated as a nuisance parameter.

4.2 Likelihood functions for binary data

4.2.1 *Log likelihood for binomial data*

Let the responses $y_1, \ldots, y_N$ correspond to independent random variables $Y_1, \ldots, Y_N$ where Y_i is assumed to be binomially distributed with index n_i and parameter Π_i. Ungrouped data are considered as a special case with $n_i = 1$ $(i = 1, \ldots, N)$. The log likelihood considered as a function of the vector $\mathbf{\Pi} = (\Pi_1, \ldots, \Pi_N)$ is

$$l(\mathbf{\Pi}; \mathbf{y}) = \sum_{i=1}^{N} \left[y_i \ln\left(\frac{\Pi_i}{1-\Pi_i}\right) + n_i \ln(1-\Pi_i) \right]. \tag{4.5}$$

The systematic part of the model specifies the relation between $\mathbf{\Pi}$ and the experimental or observational conditions as summarized by the model matrix $\mathbf{X}$ of order $N \times p$. For generalized linear models this relationship takes the form

$$g(\Pi_i) = \eta_i = \sum_j x_{ij} \beta_j; \qquad i = 1, \ldots, N,$$

so that (4.5) can be expressed as a function of the unknown parameters $\beta_1, \ldots, \beta_p$. In particular, if

$$g(\Pi) = \ln[\Pi/(1-\Pi)],$$

(4.5) becomes

$$l(\boldsymbol{\beta}; \mathbf{y}) = \sum_i \sum_j y_i x_{ij} \beta_j - \left[\sum_i n_i \ln\left(1 + \exp \sum_j x_{ij} \beta_j\right)\right], \tag{4.6}$$

where we have written $l(\boldsymbol{\beta}; \mathbf{y})$ instead of $l(\mathbf{\Pi}(\boldsymbol{\beta}); \mathbf{y})$.

The usual asymptotic results associated with statistics derived from (4.5) or (4.6) depend only on second-moment assumptions. Thus it is not essential to assume binomial variation and independence. It is sufficient to assume simply that

$$\mathrm{E}(Y_i) = n_i \Pi_i; \qquad i = 1, \ldots, N,$$

and that $$\operatorname{cov}(Y_1, \ldots, Y_N) = \operatorname{diag}[n_i \Pi_i (1-\Pi_i)].$$

In particular, the observations need not be integer-valued but we must have $0 \leqslant Y_i \leqslant n_i$.

The method of maximizing $l(\boldsymbol{\beta}; \mathbf{y})$, described in general terms in

Chapter 2, is applicable with adjusted dependent variable

$$z = \eta + (y/n - \Pi)\frac{d\eta}{d\Pi},$$

and quadratic weight function

$$W = \frac{n}{\Pi(1-\Pi)}\left(\frac{d\eta}{d\Pi}\right)^2.$$

For the logit function these simplify to

$$z = \eta + \frac{y - n\Pi}{n\Pi(1-\Pi)}$$

and $$W = n\Pi(1-\Pi).$$

The approximate covariance matrix of $\hat{\boldsymbol{\beta}}$ is $(\mathbf{X}^T\mathbf{W}\mathbf{X})^{-1}$.

4.2.2 *Quasi-likelihoods for over-dispersed data*

Data in the form of proportions frequently arise under conditions where the assumption of binomial variation is unrealistic. Usually the variance in the data is greater than that predicted by the binomial model. We refer to this phenomenon as over-dispersion or heterogeneity (Armitage, 1957; Finney, 1976; Healy, 1981; Williams, 1982). In some circumstances it is reasonable to assume

$$E(Y_i) = n_i\Pi \qquad \text{and} \qquad \text{cov}(\mathbf{Y}) = \text{diag}[\sigma^2 n_i \Pi_i(1-\Pi)], \quad (4.7)$$

where σ^2 is the dispersion parameter. The data themselves may be discrete or continuous. In (4.7) $\sigma^2 > 1$ represents over-dispersion, while $\sigma^2 < 1$ represents under-dispersion relative to the binomial model.

Quasi-likelihoods are discussed in general terms in Chapter 8. Under the second-moment assumptions (4.7) the quasi-likelihood $l(\Pi; y)$ is equal to the binomial log likelihood (4.6) so that the same estimates are obtained for $\boldsymbol{\beta}$ regardless of the value of σ^2. This is a property of the least-squares method of estimation. The asymptotic covariance matrix of $\hat{\boldsymbol{\beta}}$ is now $\sigma^2(\mathbf{X}^T\mathbf{W}\mathbf{X})^{-1}$ instead of $(\mathbf{X}^T\mathbf{W}\mathbf{X})^{-1}$. Methods of estimating σ^2 are discussed in Section 4.3. Apart from the scalar multiplier σ^2, quasi-likelihoods can be treated for the most part just like ordinary likelihoods. The principal advantage is that the second-moment assumptions on which the quasi-likelihood is based can easily be checked by graphical or other methods.

4.2.3 *The deviance function*

Let $\hat{\boldsymbol{\beta}}$ be the value of $\boldsymbol{\beta}$ obtained by maximizing the function $l(\boldsymbol{\Pi}(\boldsymbol{\beta}); \mathbf{y})$. The corresponding fitted probabilities are denoted by $\hat{\boldsymbol{\Pi}}$ and fitted counts by $\hat{\boldsymbol{\mu}}$. The deviance function $D(\mathbf{y}; \boldsymbol{\Pi})$ was introduced in Chapter 2 as a measure of the discrepancy between the observed proportions and their fitted values. For binomial data it takes the form

$$D(\mathbf{y}; \boldsymbol{\Pi}) = 2\Sigma\{y \ln(y/\hat{\mu}) + (n-y)\ln[(n-y)/(n-\hat{\mu})]\},$$

where the subscripts have been suppressed in the interests of simplicity. The same deviance function will be used whether we are dealing with genuine log likelihoods or with quasi-likelihoods.

4.3 Asymptotic theory

4.3.1 *Parameter estimation*

Parameter estimates denoted by $\hat{\boldsymbol{\beta}}$ are obtained by maximizing the log likelihood over the space specified by the systematic part of the model, (4.4), say. If the model is linear on the logistic scale the estimates may be obtained by equating the sufficient statistic, which has dimension p, to its expected value as a function of $\hat{\boldsymbol{\beta}}$. This is the basis of a method known as iterative proportional fitting and is particularly suitable if the explanatory variables are qualitative rather than quantitative. Technical details of the method can be found in Bishop *et al.* (1975).

A simple method of wider applicability is to use the Newton–Raphson method with the second-derivative matrix replaced by its expected value. As shown in Chapter 2 this amounts to weighted least squares – iterative and non-linear. In the special but important case where the systematic part is of the generalized linear type some simplification is possible along the lines described in Chapter 2. The diagonal matrix of weights for the linear logistic model is given by

$$\mathbf{W}^{-1} = \operatorname{diag}[\hat{\Pi}_i(1-\hat{\Pi}_i)/n_i],$$

and the vector of adjusted dependent variates has elements

$$z_i = \hat{\eta}_i + (y_i - n_i\hat{\Pi}_i)/[n_i\hat{\Pi}_i(1-\hat{\Pi}_i)], \tag{4.8}$$

where $\eta_i = \ln[\Pi_i/(1-\Pi_i)]$. The iterative process can therefore be written in terms of the model matrix $\mathbf{X}$ as

$$\mathbf{X}^T\mathbf{W}\mathbf{X}\hat{\boldsymbol{\beta}} = \mathbf{X}^T\mathbf{W}\mathbf{Z} \tag{4.9}$$

where $\mathbf{W}$ and $\mathbf{Z}$ are computed as functions of the current estimate of $\boldsymbol{\beta}$.

If the systematic part of the model is linear on the probit or other scale, the weighting matrix and adjusted dependent variate (4.8) must be changed accordingly; see Section 2.5. The form of the iterative equations (4.9) remains unaltered.

The convergence of process (4.9) is rarely a problem unless one or more elements of $\hat{\boldsymbol{\beta}}$ are infinite. This can occur, for example, when the data are sparse and, for some observations, $y_i = 0$ or $y_i = n_i$. Although the iterative process will not converge under these circumstances, nevertheless generally $\hat{\boldsymbol{\Pi}}^{(j)}$ tends quite rapidly toward $\hat{\boldsymbol{\Pi}}$ and the deviance tends towards its limiting value. Thus the fitted values $n_i \hat{\Pi}_i$ will be accurate but the parameter estimates may not be. The criterion used for deciding whether the process has converged should be based on $\hat{\boldsymbol{\Pi}}^{(j+1)} - \hat{\boldsymbol{\Pi}}^{(j)}$ (e.g. by using the deviance) rather than on $\hat{\boldsymbol{\beta}}^{(j+1)} - \hat{\boldsymbol{\beta}}^{(j)}$.

Some results concerning the existence and uniqueness of the parameter estimates $\hat{\boldsymbol{\beta}}$ have been given by Wedderburn (1976) and by Haberman (1977). These results show that if the link function $g(\Pi)$ is log concave, as it is for the three functions of Table 4.1, and if $0 < y_i < n_i$ for each i, then $\hat{\boldsymbol{\beta}}$ is finite and $l(\boldsymbol{\Pi}; \mathbf{y})$ has a unique maximum at $\hat{\boldsymbol{\beta}}$.

Starting values $\hat{\boldsymbol{\beta}}_0$ can be obtained as described in Chapter 2, beginning with 'fitted values' defined by $\hat{\mu} = (y + \frac{1}{2})/(n+1)$. A good choice of starting value usually reduces the number of cycles in (4.9) by about one or perhaps two. Consequently, the choice of initial estimate is usually not critical.

4.3.2 *Asymptotic theory for grouped data*

We are concerned here with the asymptotic distribution of the parameter estimates and likelihood-ratio statistics. A very careful analysis would require the consideration of a hypothetical sequence of problems as the binomial index vector $\mathbf{n}$ tends to infinity elementwise. However, we take the scalar n to represent a typical binomial index, say $n = \min(n_1, \ldots, n_N)$ such that as $n \to \infty$ each element $n_i \to \infty$

in constant proportion. The number of distinct binomial observations, N, is considered fixed.

The principal results given below refer to generalized linear models where $\mathbf{X}$, of order $N \times p$, is the model matrix and the weighting matrix $\mathbf{W}$ is diagonal with elements $(d\Pi/d\eta)^2/[n\Pi(1-\Pi)]$. The dispersion parameter σ^2, included for generality, should be set equal to 1 for binomial data. The asymptotic distribution of $\hat{\boldsymbol{\beta}}$ is given by

$$n^{\frac{1}{2}}(\hat{\boldsymbol{\beta}}-\boldsymbol{\beta}) \sim \mathrm{N}(\mathbf{0}, n\sigma^2(\mathbf{X}^{\mathrm{T}}\mathbf{W}\mathbf{X})^{-1}) + \mathrm{O}_p(n^{-\frac{1}{2}}), \qquad (4.10)$$

assuming, as usual, the adequacy of the model. The error term in (4.10) means that the cumulative distribution of $n^{\frac{1}{2}}(\hat{\boldsymbol{\beta}}-\boldsymbol{\beta})$ differs from the cumulative p-variate normal distribution by a term of order $n^{-\frac{1}{2}}$. In other words probability calculations based on the Normal approximation (4.10) have an error of order $n^{-\frac{1}{2}}$. It is also possible to show that the bias in $\hat{\boldsymbol{\beta}}$ is $\mathrm{O}(n^{-1})$.

Several purely technical details have been ignored in the development leading to (4.10). An important requirement is that p, the number of parameters, should remain fixed as $n \to \infty$. In practice this means that p should not be a large fraction of N, particularly if some of the binomial denominators are small.

The second major asymptotic result concerns the distribution of the deviance, $D(\mathbf{Y};\hat{\mathbf{\Pi}})$. The null distribution is given by

$$D(\mathbf{Y};\hat{\mathbf{\Pi}}) \sim \sigma^2\chi^2_{N-p} + \mathrm{O}_p(n^{-\frac{1}{2}}), \qquad (4.11)$$

where as before $\sigma^2 = 1$ for binomial data.

Results (4.10) and (4.11) can be deduced from the general theory of Chapter 8.

For over-dispersed data it is necessary to find an estimate of σ^2 in order to use (4.10). There are several possibilities here, including the estimator $D(\mathbf{Y};\hat{\mathbf{\Pi}})/(N-p)$ based on the deviance. Consistency as $n \to \infty$ follows from (4.11). However, we prefer the estimator

$$\begin{aligned}\tilde{\sigma}^2 &= \frac{1}{N-p}\sum_{i=1}^{N}(y_i - n_i\hat{\Pi}_i)^2/[n_i\hat{\Pi}_i(1-\hat{\Pi}_i)] \\ &= X^2/(N-p),\end{aligned}$$

where X^2 is Pearson's chi-squared statistic. The main reason for preferring $\tilde{\sigma}^2$ is that $\tilde{\sigma}^2$ is consistent in the limit as $N \to \infty$ with n fixed and its asymptotic distribution as $n \to \infty$ is known to be

$\sigma^2\chi^2_{N-p}/(N-p)+O_p(n^{-\frac{1}{2}})$. Thus $\tilde{\sigma}^2$ is a satisfactory estimator in either limit. Furthermore it can be shown that as $n\to\infty$, $\hat{\boldsymbol{\beta}}$ and $\tilde{\sigma}^2$ are asymptotically independent.

Approximate confidence intervals for an element, β_1 say, of $\boldsymbol{\beta}$ are formed in the usual way, namely

$$\hat{\beta}_1 \pm k_{\alpha/2}(i^{(11)})^{\frac{1}{2}} \tag{4.12}$$

if $\sigma^2 = 1$ is known and

$$\hat{\beta}_1 \pm \tilde{\sigma} t_{\alpha/2,\, N-p}(i^{(11)})^{\frac{1}{2}} \tag{4.13}$$

if σ^2 is unknown. Here $i^{(11)}$ is the (1, 1) element of $(\mathbf{X}^T\mathbf{W}\mathbf{X})^{-1}$, $\Phi(k_\alpha) = 1-\alpha$ and $t_{\alpha/2,\, N-p}$ is the $100(1-\alpha/2)$ percentage point of the t-distribution with $N-p$ degrees of freedom. The intervals given above have coverage probability $1-\alpha+O(n^{-\frac{1}{2}})$, the probability in each tail being $\alpha/2+O(n^{-\frac{1}{2}})$.

Approximate confidence intervals can also be based on the likelihood-ratio statistic or on the score statistic (see e.g. Cox and Hinkley, 1974, ch. 9). Intervals based on these methods have the strong conceptual advantage of being invariant under reparameterization, in contrast to (4.12) and (4.13). However, they are computationally much more cumbersome, requiring, in general, numerous maximizations of the likelihood function.

4.3.3 *Asymptotic theory for ungrouped data*

In this section we consider asymptotic approximations appropriate as $N\to\infty$ but where the binomial totals n_i remain small. The general theory of Chapter 8 shows that $\hat{\boldsymbol{\beta}}$ is consistent and asymptotically Normal in this limit. The only difficulty concerns the estimation of σ^2 where over- or under-dispersion is suspected. For known σ^2 all the results of the previous section apply, the error being of order $N^{-\frac{1}{2}}$ rather than $n^{-\frac{1}{2}}$.

Of the two estimators for σ^2 given in the previous section only $\tilde{\sigma}^2$ is consistent as $N\to\infty$ with $\{n_i\}$ bounded. It is readily shown that

$$E(\tilde{\sigma}^2) = \sigma^2 + O(N^{-1})$$

and

$$\text{var}(\tilde{\sigma}^2) = 2\sigma^4(1+\tfrac{1}{2}\bar{\gamma}_2)/(N-p)+O(N^{-2}), \tag{4.14}$$

where $\bar{\gamma}_2$ is the average of the standardized fourth cumulants or the average standardized kurtosis of the observations. Thus the effective degrees of freedom associated with $\tilde{\sigma}^2$ are

$$(N-p)/(1+\tfrac{1}{2}\bar{\gamma}_2).$$

The result just given has a number of ready applications provided that some reasonable estimate can be obtained for $\bar{\gamma}_2$. Denote by f the effective degrees of freedom or an estimate thereof; the interval (4.13) then becomes

$$\hat{\beta}_1 \pm \tilde{\sigma} t_{\alpha/2, f} (i^{(11)})^{\frac{1}{2}}. \tag{4.15}$$

For the binomial distribution

$$\gamma_2 = [1-6\Pi(1-\Pi)]/[n\Pi(1-\Pi)]$$

and has a minimum of $-2/n$ at $\Pi = 0.5$, becomes zero at $\Pi = 0.2113$ and 0.7887, and increases without limit as $\Pi \to 0$ or $\Pi \to 1$. For our purposes, it is not necessary that $\bar{\gamma}_2$ be estimated with great precision: the interval (4.15) depends only weakly on f provided f is large. One simple and obvious solution, at least for distributions near the binomial, is to take

$$\bar{\gamma}_2 = \frac{1}{N} \sum_i [1-6\hat{\Pi}_i(1-\hat{\Pi}_i)]/[n_i \hat{\Pi}_i(1-\hat{\Pi}_i)] \tag{4.16}$$

although there are clearly other possibilities.

The computations just given show that, when setting confidence limits, we should err on the side of caution particularly if many of the Π_is lie outside the range (0.2, 0.8). For example, if all the Πs are in the vicinity of 0.1 and $n_i = 1$ for all i, then the effective degrees of freedom are $0.28 \times (N-p)$. For $n_i = 4$, for all i, the fraction increases to 0.61. The larger the value of $\{n_i\}$ the closer the results of this section approach those of Section 4.3.2.

4.3.4 *Residuals and goodness-of-fit tests*

There are two principal approaches to the assessment of goodness of fit. The method most commonly employed in the literature on discrete data involves computing a summary statistic such as the deviance or Pearson's chi-squared statistic. For grouped data the null distribution of these statistics is asymptotically chi-squared. We shall use these summary statistics on occasion but concentrate mainly on an examination of the residuals. When summary statistics are required it is often helpful to improve the usual chi-squared approximation by means of a first-order correction term. For example the mean of the deviance assuming binomial variation is, from Appendix B,

$$\mathrm{E}(D(\mathbf{Y};\hat{\mathbf{\Pi}})) = N-p+\Sigma[1-\Pi(1-\Pi)]/[6n\Pi(1-\Pi)]+\mathrm{O}(n^{-2}). \tag{4.17}$$

On the right of (4.17) Π_i can be replaced by $\hat{\Pi}_i$. In other words the mean of $D(\mathbf{Y};\hat{\mathbf{\Pi}})$ differs from the reference chi-squared distribution by the factor $1+c$, where

$$c = \frac{1}{N-p}\sum_i [1-\Pi_i(1-\Pi_i)]/[6n_i\,\Pi_i(1-\Pi_i)]. \tag{4.18}$$

Lawley (1956) showed that all moments of the adjusted statistic $D(\mathbf{y};\hat{\mathbf{\Pi}})/(1+c)$ differ from those of χ^2_{N-p} by terms $\mathrm{O}(n^{-2})$. Here N is taken as fixed and $n\to\infty$. This adjustment improves probability calculations provided that $n\Pi_i$ and $n(1-\Pi_i)$ both exceed 1.

The approximation to Pearson's statistic can be improved by taking account of the effective degrees of freedom. In other words we use X^2 as summary statistic, the reference distribution being χ^2_f where

$$f = (N-p)/(1+\tfrac{1}{2}\bar{\gamma}_2).$$

The two summary statistics just given are useful when the assumption of binomial variation is not questioned. The reference chi-squared distributions do not apply to over-dispersed data. A more important drawback is that neither summary statistic gives any indication of the direction of departures from the fitted model. To investigate this we need to examine residuals. Usually the residuals are examined graphically though it is not always necessary to construct a graph, particularly when N is small.

The simplest, most direct and probably most useful definition of residual in the context of binomial or binomial-like data is the Pearson residual, as defined in Section 2.4,

$$(y_i-n_i\,\hat{\Pi}_i)/[n_i\,\hat{\Pi}_i(1-\hat{\Pi}_i)]^{\frac{1}{2}}. \tag{4.19}$$

If n_i is small or Π_i is near 0 or 1, the adjusted deviance residual may be found preferable. It is defined by

$$\begin{aligned} d_i = \pm\{2y_i\ln(y_i/n\hat{\Pi}_i)+2(n_i-y_i)\ln[(n_i-y_i)/(n_i-n_i\,\hat{\Pi}_i)]\}^{\frac{1}{2}} \\ +(2\hat{\Pi}_i-1)/\{6[n_i\,\hat{\Pi}_i(1-\hat{\Pi}_i)]^{\frac{1}{2}}\}, \end{aligned} \tag{4.20}$$

where the sign is that of $y_i-n_i\,\hat{\Pi}_i$. The adjusted deviance residuals are closely related to the components of the likelihood-ratio statistic and are more nearly Normally distributed than the Pearson residuals. In extreme cases, if many or all of the n_i are equal to one, it is best to group the observations before computing residuals. Procedures for model checking using the residuals are discussed in Chapter 11.

In constructing the residual-like quantities (4.19) and (4.20) we have ignored the variation in the estimate of Π. It is possible to standardize further. To the first order the covariance matrix of $(y_i/n_i - \hat{\Pi}_i)$ is $\mathbf{V} - \mathbf{X}(\mathbf{X}^T\mathbf{V}^-\mathbf{X})^-\mathbf{X}^T$ where $\mathbf{V} = \text{diag}[\Pi_i(1-\Pi_i)/n_i]$ and $\mathbf{X}$ is the model matrix. Let the diagonal elements of this matrix be $\Pi_i(1-\Pi_i)(1-c_{ii})/n_i$. Thus to a closer order of approximation $r_i/(1-c_{ii})^{\frac{1}{2}}$ and $d_i/(1-c_{ii})^{\frac{1}{2}}$ are approximately N(0, 1). The analysis given here ignores covariances among the residuals: this is not easily justified particularly if p is a substantial fraction of N, but it is difficult to construct a general method for dealing with dependencies of this nature.

* 4.4 Conditional likelihoods

The asymptotic results given in Section 4.3.3 do not apply when the number of parameters is comparable in magnitude to N, the quantity tending to infinity. When this happens many of the parameters are often of the type called incidental or nuisance parameters; usually only a few of the parameters measure quantities of interest. In some circumstances the nuisance parameters can be eliminated by appropriate conditioning and, in principle, uniquely defined exact inference follows. Here we consider approximate methods based on appropriate conditional likelihood functions.

4.4.1 *Conditional likelihood for 2 × 2 tables*

We consider first a single 2×2 table or, equivalently, a pair of independent binomial observations Y, Z with denominators n_1, n_2 and parameters Π_1 and Π_2. Let

$$\psi = \Pi_1(1-\Pi_2)/[\Pi_2(1-\Pi_1)],$$

the odds ratio (sometimes also referred to as the relative risk), be the parameter of interest. Then the conditional distribution of Y given $Y+Z = m$, say, is non-central hypergeometric, depends only on ψ

and is given by

$$\mathrm{pr}(Y = y|n_1, n_2, m) = \binom{n_1}{y}\binom{n_2}{m-y}\psi^y \Big/ \sum_s \binom{n_i}{s}\binom{n_2}{m-s}\psi^s, \quad (4.21)$$

where the sum in the denominator extends from $s = \max(0, m-n_2)$ to $\min(n_1, m)$. Denote by $P_r(\psi)$ the polynomial

$$P_r(\psi) = \sum_s s^r \binom{n_1}{s}\binom{n_2}{m-s}\psi^s,$$

so that the conditional moments of Y are

$$\mathrm{E}(Y^r) = P_r(\psi)/P_0(\psi); \qquad r = 1, 2, \ldots. \quad (4.22)$$

The conditional log likelihood for ψ is

$$y \ln \psi - \ln P_0(\psi), \quad (4.23)$$

which could, though with some difficulty, be written as a function of the conditional mean $\mu = \mathrm{E}(Y) = P_1(\psi)/P_0(\psi)$. Suppose now that other data sets with the same classifying factors are available in the form of several 2×2 tables; the parameter ψ might be common to all tables or it might vary in a systematic way from table to table. Usually the different tables correspond to the stratification of a population by some quantitative variable x, possibly vector-valued. For the moment we take $\boldsymbol{\psi}$ to be a vector of length N restricted only in that its elements are positive.

The full log likelihood considered as a function of $\boldsymbol{\psi}$ is the sum of contributions each having the form (4.23). Thus

$$l(\boldsymbol{\psi}; \mathbf{y}) = \Sigma y_i \ln \psi_i - \Sigma \ln P_0^{(i)}(\psi_i), \quad (4.24)$$

where the superscript in $P_0^{(i)}(\psi)$ is included to emphasize that the coefficients and the degree of this polynomial depend on the conditioning quantities in the ith table. The systematic part of the model describing the relation between the vector of odds ratios $\boldsymbol{\psi}$ and the model matrix $\mathbf{X}$ can have any arbitrary form. In applications multiplicative models for the odds ratios are usually most plausible. Thus we put $\theta_i = \log \psi_i$ and consider linear models for the N-vector $\boldsymbol{\theta}$, i.e.

$$\boldsymbol{\theta} = \mathbf{X}\boldsymbol{\beta} \quad (4.25)$$

in the usual notation. This model is linear on the canonical scale and the log likelihood (4.24) reduces to

$$\boldsymbol{\beta}^{\mathrm{T}}\mathbf{X}^{\mathrm{T}}\mathbf{Y} - \sum_i \ln[P_0^{(i)}(\mathrm{e}^{\theta_i})], \quad (4.26)$$

so that the sufficient statistic $\mathbf{X}^T\mathbf{Y}$ is of dimension p. Thus, in principle at least, exact tests for $\boldsymbol{\beta}$ or components thereof can be constructed via the appropriate conditional distribution; however, for reasons of simplicity we concentrate on approximate methods based on the maximum conditional likelihood estimate.

4.4.2 *Parameter estimation*

Let $\boldsymbol{\mu}$ and $\mathbf{V}$ be the conditional mean vector and covariance matrix of $\mathbf{Y}$ computed from the conditional distribution (4.21). Differentiating (4.26), the equations for $\hat{\boldsymbol{\beta}}$ can be written

$$\mathbf{X}^T\mathbf{y} = \mathrm{E}(\mathbf{X}^T\mathbf{Y}; \hat{\boldsymbol{\beta}}).$$

In other words we choose $\hat{\boldsymbol{\beta}}$ by equating the sufficient statistic to its expected value. The Newton–Raphson method for solving this system (Section 2.5) can be written

$$\hat{\boldsymbol{\beta}} = \hat{\boldsymbol{\beta}}_0 + (\mathbf{X}^T\mathbf{V}\mathbf{X})^{-1}\mathbf{X}^T(\mathbf{y} - \hat{\boldsymbol{\mu}}_0), \tag{4.27}$$

where the quantities on the right are computed at $\hat{\boldsymbol{\beta}}_0$. Convergence is guaranteed provided that $\hat{\boldsymbol{\beta}}$ is finite. The principal difficulty with (4.27) is that the elements of $\mathbf{V}$ and $\boldsymbol{\mu}$ involve polynomials $P_0^{(i)}(\psi)$, $P_1^{(i)}(\psi)$ and $P_2^{(i)}(\psi)$ whose degree depends on the marginal totals in each table. These can be arbitrarily large. There are $3N$ polynomials in all, each of which must be recomputed at each cycle. Thus the apparent simplicity of (4.27) belies the computational complexity of the components of $\mathbf{V}$ and $\boldsymbol{\mu}$.

In the simplest of cases where we have several tables each with a common value of the odds ratio, the initial adjustment in (4.27) beginning with $\hat{\boldsymbol{\beta}}_0 = \mathbf{0}$ has the form $[T - \mathrm{E}(T)]/\mathrm{var}(T)$ where $T = \Sigma Y$. This is closely related to the Mantel–Haenszel test (Mantel and Haenszel, 1959) which, when used with a continuity correction, may be written

$$[|T - \mathrm{E}(T)| - 1/2]^2/\mathrm{var}(T).$$

In fact the Mantel–Haenszel test is simply the score test based on the derivative of the log likelihood (4.24) (see, for example, Day and Byar, 1979; Pregibon, 1982).

The following numerical method is suggested as a means of approximating $\boldsymbol{\mu}$ and $\mathbf{V}$ in (4.27). Consider the distribution (4.21) with conditional mean μ and conditional variance v. Then it is easily shown (Mantel and Hankey, 1975) that

$$\frac{\mu(n_2-m+\mu)+v}{(n_1-\mu)(m-\mu)+v} = \psi. \tag{4.28}$$

Considering v as fixed, i.e. not a function of ψ, we obtain from (4.28) a quadratic equation for the conditional mean, μ:

$$(\psi-1)\mu^2 - \mu[n_1+n_2+(m+n_1)(\psi-1)] + mn_1\psi + v(\psi-1) = 0. \tag{4.29}$$

If v were known, (4.29) could be used to compute the exact conditional mean, thus avoiding the computation of the polynomials $P_0(\psi)$, $P_1(\psi)$ and $P_2(\psi)$. Fortunately the solution to (4.29) depends only weakly on v so that any reasonable estimate will suffice. In fact if we set $v = 0$ the solution to (4.29) gives the unconditional mean. In practice we choose as an initial estimate for v the conditional variance of Y when $\psi = 1$. This gives

$$v_0 = n_1 n_2 m(n_1+n_2-m)/[(n_1+n_2)^2(n_1+n_2-1)]. \tag{4.30}$$

In subsequent iterations the current estimate for μ is available so that the approximation

$$v \simeq \left(\frac{n_1+n_2}{n_1+n_2-1}\right)\left(\frac{1}{\mu}+\frac{1}{n_1-\mu}+\frac{1}{m-\mu}+\frac{1}{n_2-m+\mu}\right)^{-1} \tag{4.31}$$

can be used in place of (4.30). Equation (4.31) is the leading term in an asymptotic expansion for v and is therefore consistent.

The numerical approximation just described gives very satisfactory results provided that each of the four marginal totals n_1, n_2, m, n_1+n_2-m in each table exceeds about 5. In fact the method gives exact results for $n_1 = n_2 = 1$, corresponding to matched pairs, but the rate of convergence may be slow. If the marginal totals are less than 5 the exact calculations are not onerous and we would recommend them.

An alternative algorithm for computing the exact conditional maximum-likelihood estimate with minimum arithmetic is described by Gail *et al.* (1981). Their exact method is more widely applicable than the approximate method described here.

The estimate of $\boldsymbol{\beta}$ obtained from (4.27) is consistent and asymptotically Normal in the limit as $N\to\infty$ or, for fixed N, as the marginal totals for each table become large. A further by-product of the method of estimation is that $(\mathbf{X}^T\mathbf{V}\mathbf{X})^{-1}$ is a consistent estimate of cov $(\hat{\boldsymbol{\beta}})$ in either of the limits described below.

4.4.3 *Combination of information from several 2 × 2 tables*

The description given in the previous section applies to arbitrary linear regression models for the logarithm of the odds ratio. In many applications, the objective is to combine information concerning the odds ratio from several tables, it being assumed that the value of ψ is the same in each table. This corresponds to a linear model in which the matrix **X** of (4.25) consists of a single column of 1s, reflecting the fact that ψ is assumed constant over tables. Estimation proceeds as described in (4.27).

It is important to emphasize here that we cannot estimate ψ simply by referring to the 2 × 2 table of totals. The well known paradox due to E. H. Simpson (1951) shows that the odds ratio in the table of totals may be quite different from the odds ratio of the constituent tables.

In a pioneering paper on retrospective studies, Mantel and Haenszel (1959) gave an estimator of the odds ratio that avoids Simpson's paradox and is different from the conditional estimator. The principal advantage of their estimator is that it can be computed non-iteratively. It is convenient at this stage to make a temporary change of notation so that the elements in a 2 × 2 table are as follows:

			Total
	$y \equiv y_{00}$	y_{01}	n_1
	y_{10}	y_{11}	n_2
Total	m	$n-m$	n

The Mantel–Haenszel estimator of ψ from several such tables is given by

$$\hat{\psi}_{\mathrm{MH}} = \frac{\Sigma\,(y_{00}\,y_{11}/n)}{\Sigma\,(y_{01}\,y_{10}/n)},$$

where the subscript i referring to tables has been dropped. The probability limit of the numerator in an obvious notation in $\Sigma[(\mu_{00}\,\mu_{11}+v)/n]$ while that of the denominator is $\Sigma[(\mu_{01}\,\mu_{10}+v)/n]$. Consistency follows from the fact that

$$\psi = (\mu_{00}\,\mu_{11}+v)/(\mu_{01}\,\mu_{10}+v).$$

The asymptotic variance is more difficult to compute but shows that $\hat{\psi}_{\mathrm{MH}}$ is generally slightly less efficient than the conditional maximum-likelihood estimate.

Significance tests of $H_0: \psi = 0$ are based on the sufficient statistic $T = \Sigma Y_{00} \equiv \Sigma Y$, the sum of the (0, 0) elements in each of the tables. The exact null mean and variance of T are given by

$$\Sigma \mu = \Sigma (mn_1/n)$$

and

$$\Sigma V = \Sigma n_1 n_2 m(n-m)/[n^2(n-1)].$$

In principle the exact null distribution of T could be computed. In practice the Normal approximation with continuity correction is usually considered adequate. The approximate two-sided significance level can be found by referring

$$(|T - \Sigma \mu| - \tfrac{1}{2})^2 / \Sigma V \tag{4.32}$$

to the χ_1^2 distribution.

4.4.4 *Residuals*

Here, as in Section 4.3.4, a number of definitions can be used for residuals. For simplicity we use

$$(y_i - \hat{\mu}_i)/\hat{V}_i^{\frac{1}{2}}. \tag{4.33}$$

All the relevant quantities in (4.33) arise naturally in the fitting procedure. Residuals based on the likelihood-ratio statistic have better distributional properties than (4.33) but are tedious to compute. In many applications the number of strata N is not sufficiently large to warrant plotting the residuals against other quantities and visual inspection is adequate. When N is sufficiently large a plot of r_i against $\hat{\theta}_i$ or $\hat{\psi}_i$ could be used to diagnose deviations from the assumed linear relation (4.25). Plots of the residuals against $\hat{\mu}_i$ are not so readily interpretable because of the varying marginal totals.

4.5 Examples

4.5.1 *Habitat preferences of lizards*

The following data are in many ways typical of social-science investigations, although the example concerns the behaviour of lizards rather than humans. The data, taken from Schoener (1970), have subsequently been analysed by Fienberg (1970b) and by Bishop *et al.* (1975). The data, which concern two species of lizard, *Grahami* and *Opalinus*, were collected by observing occupied sites or perches and recording the appropriate description, namely species involved, time of day, height and diameter of perch and whether the site was sunny or shaded. Here, time of day is recorded as early, mid-day or late.

As often with such problems, several analyses are possible depending on the purpose of the investigation. We might, for example, wish to compare how preferences for the various perches vary with the time of day regardless of the species involved. We find, for example, by inspection of the data (Table 4.2), that shady sites are preferred to sunny sites at all times of day but particularly so at mid-day. Furthermore, again by inspection, low perches are preferred to high ones and small-diameter perches to large ones. There is, of course, the possibility that these conclusions are an artefact of the data-collection process and that, for example, occupied sites at eye level or below are easier to spot than occupied perches higher up. In fact selection bias of this type seems inevitable unless some considerable effort is devoted to observing all lizards in a given area.

Table 4.2 *A comparison of site preferences for two species of lizard,* Grahami *and* Opalinus

			T								
			Early			*Mid-day*			*Late*		
S	*D* (in)	*H* (ft)	*G*	*O*	*Total*	*G*	*O*	*Total*	*G*	*O*	*Total*
Sun	$\leqslant 2$	< 5	20	2	22	8	1	9	4	4	8
		$\geqslant 5$	13	0	13	8	0	8	12	0	12
	> 2	< 5	8	3	11	4	1	5	5	3	8
		$\geqslant 5$	6	0	6	0	0	0	1	1	2
Shade	$\leqslant 2$	< 5	34	11	45	69	20	89	18	10	28
		$\geqslant 5$	31	5	36	55	4	59	13	3	16
	> 2	< 5	17	15	32	60	32	92	8	8	16
		$\geqslant 5$	12	1	13	21	5	26	4	4	8

H, height; *D*, diameter; *S*, sunny/shady; *T*, time of day; *G*, *Grahami*; *O*, *Opalinus*.

A similar analysis, but with the same deficiencies, can be made for each species separately.

Suppose instead that an occupied site, regardless of its position, diameter and so on, is equally difficult to spot whether occupied by

Grahami or *Opalinus*. This would be plausible if the two species were similar in size and colour. Suppose now that the purpose of investigation is to compare the two species with regard to their preferred perches. Thus we see, for example, that of the 22 occupied perches of small diameter low on the tree observed in a sunny location early in the day, only two, or 9%, were occupied by *Opalinus* lizards. For similar perches observed later in the day the proportion is four out of eight, i.e. 50%. On this comparison, therefore, it appears that, relative to *Opalinus*, *Grahami* lizards prefer to sun themselves early in the day.

To pursue this analysis more formally we take as fixed the total number n_{ijkl} of occupied sites observed for each combination of i = perch height, j = perch diameter, k = sunny/shady and l = time of day. In the language of statistical theory, these totals are ancillary provided that the purpose of the investigation is to compare preferences of *Grahami* and *Opalinus*. The response variable y_{ijkl} gives the observed number or, equivalently, the proportion of the n_{ijkl} observed sites occupied by *Grahami* lizards. Equivalently we could work with $n_{ijkl}-y_{ijkl}$, the number of sites occupied by *Opalinus* lizards. We take the variable Y_{ijkl} to be binomially distributed with index n_{ijkl} and parameter Π_{ijkl} where Π is the probability that an occupied site is in fact occupied by a *Grahami* lizard. Of course the possibility of over-dispersion relative to the binomial model must be borne in mind.

At the exploratory stage, probably the simplest analysis of these data is obtained by transforming to the logistic scale. To reduce bias and to avoid problems associated with infinite values we add 1/2 to both numerator and denominator in the transformation (Cox, 1970, p. 35). Thus $z_1 = \ln(20.5/2.5) = 2.1041$ with approximate variance $(20.5)^{-1}+(2.5)^{-1} = 0.4488$. This kind of analysis is usually satisfactory if all the counts are moderately large. In this particular example they are not and so it will be necessary to confirm our findings by a different method. To maintain balance, the observation (0, 0) is transformed to $z_{12} = 0.0$ with 'variance' 4.0.

In Table 4.3 the transformed data with their variances are listed in standard order corresponding to the four factors *H*, *D*, *S* and *T*, and the four steps of Yates's algorithm are performed to obtain the appropriate raw contrasts associated with the four factors, height, diameter, sunny/shady and time of day, for the last of which linear (L) and quadratic (Q) contrasts are used. Variances are computed in

Table 4.3 *Computation of logistic factorial standardized contrasts for lizard data*

Transformed value	Estimated variance	Raw contrast	Estimated variance	Parameter	Absolute standardized contrast
2.1041	0.4488	30.9091	20.16	I	—
3.2958	2.0741	11.0985	20.16	H	2.47
0.8873	0.4034	−12.4507	20.16	D	2.77
2.5649	2.1538	−5.3629	20.16	HD	1.19
1.0986	0.1159	−5.4697	20.16	S	1.22
1.7451	0.2136	−0.1739	20.16	HS	0.04
0.1214	0.1217	3.9169	20.16	DS	0.87
2.1203	0.7467	5.4015	20.16	HDS	1.20
1.7346	0.7843	−8.3504	11.79	T_L	2.43
2.8332	2.1176	−1.9644	11.79	HT_L	0.57
1.0986	0.8889	−2.1334	11.79	DT_L	0.62
0.0000	4.0000	−6.2926	11.79	HDT_L	1.83
1.2209	0.0632	2.0120	11.79	ST_L	0.59
2.5123	0.2402	−1.7596	11.79	HST_L	0.51
0.6214	0.0473	−0.4950	11.79	DST_L	0.14
1.3633	0.2283	2.0210	11.79	$HDST_L$	0.59
0.0000	0.4444	−3.2438	45.27	T_Q	0.48
3.2189	2.0900	4.9986	45.27	HT_Q	0.74
0.4520	0.4675	3.2024	45.27	DT_Q	0.48
0.0000	1.3333	2.8772	45.27	HDT_Q	0.43
0.5664	0.1493	−5.6242	45.27	ST_Q	0.84
1.3499	0.3598	−6.2738	45.27	HST_Q	0.93
0.0000	0.2353	−1.2542	45.27	DST_Q	0.19
0.0000	0.4444	0.4584	45.27	$HDST_Q$	0.07

a similar way, the coefficients being squared. Thus all main effects and interactions involving H, D and S only have the same variance, 20.16. Similarly for terms involving T_L and for terms involving T_Q. Finally we compute the standardized contrasts: of these, only the main effects of H and D and the linear effect of time, with standardized contrasts in excess of 2.4, appear significant.

Because of the many small observed counts in this particular example, some caution is required in interpreting the contrasts in Table 4.3. It is possible, for example, that the addition of 1/2 to each count before transforming could swamp the data. Indeed this appears to have happened for the sunny/shady contrasts. Here few *Opalinus* lizards were observed in sunny locations, so that the

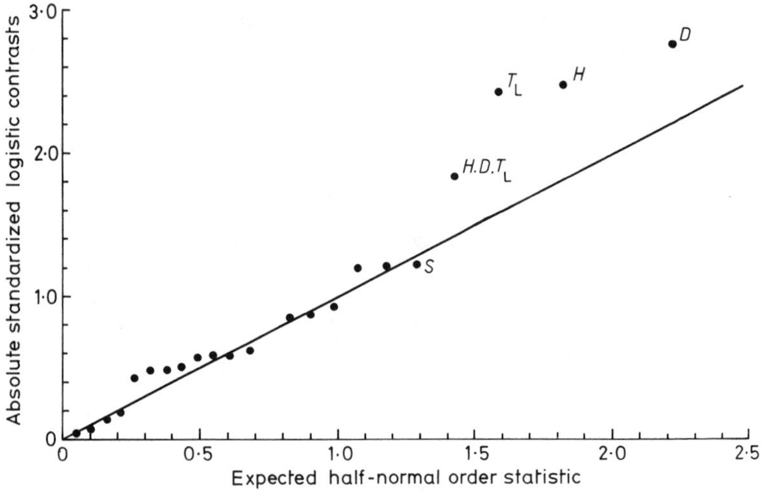

Fig. 4.1. *Half-normal plot of absolute standardized logistic contrasts for the lizard data.*

addition of 1/2 to each of these counts has a marked effect on the S contrast, reducing it towards zero.

The half-normal plot (Daniel, 1959) of the ordered absolute standardized logistic contrasts against their expected values (Fig. 4.1) suggests that the main effects of height and diameter and the linear effect of time are significant though not overwhelmingly so. The three-factor interaction $H \cdot D \cdot T_L$ also deviates from the theoretical line but this is an aberration well within the sampling limits especially when an allowance is made for selection. No allowance for selection would normally be made for the main effects.

We now consider an alternative analysis by maximum likelihood. The preceding analysis in Table 4.3 and Fig. 4.1 suggests that the structure of these data is fairly simple; there appear to be no significant interactions. We are thus led to consider the linear logistic model including all four main effects, H, D, S and T. The model can be written symbolically as $H+D+S+T$ or, in subscript notation,

$$\text{logit}(\Pi_{ijkl}) = \mu + \alpha_i + \beta_j + \gamma_k + \delta_l, \tag{4.34}$$

where α, β, γ and δ refer to the four factors H, D, S and T. In fact this model fits the data quite well with no evidence of over-dispersion. All main effects including S are significant at the 5% level, illustrating

Table 4.4 *Examination of two-factor interactions for lizard data*

*Model description**	*Degrees of freedom*	*Deviance*	*First difference*
Main effects only	17†	14.20	
Main + $T \cdot S$	15	12.93	1.27
Main + $T \cdot H$	15	13.68	0.34
Main + $T \cdot D$	15	14.16	0.04
Main + $S \cdot H$	16	11.98	2.22
Main + $S \cdot D$	16	14.13	0.07
Main + $H \cdot D$	16	13.92	0.28

* The factors here are time of day (T), sunny/shady (S), height (H) and diameter (D).

† Degrees of freedom reduced by one because no occupied sites were observed for $(i, j, k, l) = (2, 2, 2, 2)$.

Table 4.5 *Parameter estimates for the linear logistic model (4.34)*

Parameter	*Estimate*	*SE*
μ	1.945	0.34
H, height > 5 ft	1.13	0.26
D, diameter > 2 in	−0.76	0.21
S, shady	−0.85	0.32
$T(2)$, mid-day	0.23	0.25
$T(3)$, late	−0.74	0.30

the drawbacks of the previous analysis where the effect of S appeared to be insignificant. None of the two-factor interactions appears significant; the relevant statistics are given in Table 4.4. Parameter estimates associated with model (4.34) are given in Table 4.5 where we use the convention of setting the first level of each factor to zero. It would be possible here to replace T by single contrast corresponding to late afternoon versus earlier in the day without unduly affecting the fit. This would reduce the number of parameters by one but would not greatly simplify the statement of conclusions.

Finally, a rather informal examination of the standardized residuals reveals no unexpected features or patterns.

The principal conclusions to be drawn are as follows. High perches are more likely to be occupied by *Grahami* lizards than are low perches. The ratio of the odds for high versus low perches is $3.10 = \exp(1.13)$ and this ratio applies under all conditions of shade, perch diameter and time of day. It would be false to conclude from this that *Grahami* lizards prefer high perches to low perches. We may however conclude that *Grahami* lizards have less aversion to high perches than do *Opalinus* lizards, so that an occupied high perch is more likely to contain a *Grahami* than is an occupied low perch.

Similar conclusions may be drawn regarding the effects of shade, perch diameter and time of day on the probability that an occupied site contains a *Grahami* lizard. The odds are largest for small-diameter lofty perches in a sunny location at mid-day (or in the morning). In fact only *Grahami* and no *Opalinus* lizards were observed under these conditions. Because there is no interaction, the odds are smallest for the converse factor combinations.

These conclusions differ from those of Fienberg (1970b) and Bishop *et al.* (1975) who found an interaction between H and D and between S and T regarding their effect on species' preferences. The principal reason for this difference appears to be the fact that these authors considered several issues simultaneously and did not condition on the totals n_{ijkl}, which are regarded as ancillary in the analysis given here.

4.5.2 *Effect of a drug on cardiac deaths*

This example concerns the estimation of the odds ratio in a 2×2 table and the computation of significance levels. The data are taken from a study of the effect of the drug sulphinpyrazone on cardiac death after myocardial infarction (Anturane Reinfarction Trial Research Group 1978, 1980).

In keeping with well established procedures, especially in cancer studies, the primary analysis given here relates to deaths from all causes during the course of the trial. In fact the numbers in Table 4.6 are large enough for an unconditional analysis to have been considered satisfactory. Take Y_1 and Y_2 to be the numbers of deaths in the drug group and in the control group respectively. The corresponding estimated odds are $41/692$ and $60/682$ giving an odds ratio of $\hat{\psi} = 41 \times 682/(60 \times 692) = 0.6735$ or $\ln \hat{\psi} = -0.3953$. The approximate variance of $\log \hat{\psi}$ is the sum of reciprocals $1/41 + 1/60 + 1/692 + 1/682$, giving an approximate standard error of 0.2097 or an approximate standardized statistic of -1.88. This corresponds to a two-sided significance level of 6% which, while strongly suggestive of a treatment effect, is hardly conclusive.

In the conditional analysis described in Section 4.4 we regard the

Table 4.6 *A study of the effect of the drug sulphinpyrazone*

	Deaths from all causes	*Survivors*	*Total*
Sulphinpyrazone	41	692	733
Placebo	60	682	742
Total	101	1374	1475

margins as fixed. The exact one-sided significance level for the hypothesis of no treatment effect is the conditional probability

$$p_{\text{obs}}^{-} = \text{pr}\,(Y_1 \leqslant 41 \,|\, \text{margins}, \psi = 1).$$

This is known as Fisher's exact test: the required probability $p_{\text{obs}}^{-} = 3.63\%$ can be obtained from the tail area of the central hypergeometric distribution. The two-sided significance level is usually found by doubling the one-sided value, although this is not the only possibility (Cox and Hinkley, 1974, p. 119). Alternatively, an approximate significance level can be found based on the continuity-corrected chi-squared test (Fleiss, 1981, p. 26) which is equivalent to using the statistic (4.32).

The conditional estimate $\hat{\psi}_c$ is found using (4.27) where $\beta = \log \psi$. The procedure converges to $\log \hat{\psi}_c = -0.3951$ and $\text{var}\,(\log \hat{\psi}) \simeq V^{-1} = 0.0439$ or $\text{SE}(\log \hat{\psi}_c) = 0.2096$, in good agreement with the unconditional analysis.

While the conditional analysis would be preferred on general principle, in applications it would generally be employed only where the observed counts were small. In the present example the unconditional analysis gives perfectly satisfactory answers.

To summarize, the estimated effect of the drug is to decrease the odds of death over the trial period by a factor of $\hat{\psi} = 1.48$. This is a large effect, and medically, if not statistically, significant. However, at the 5% level of significance the data are consistent with a wide range of ψ-values including the null value $\psi = 1$. Thus, on the analysis given here, the evidence for a beneficial effect of the drug is not overwhelming.

A further example involving linear regression of the log odds ratio is given in Chapter 8.

4.6 Bibliographic notes

The literature on the analysis of discrete data generally, and binary data in particular, is very extensive. No attempt is made here to provide an exhaustive list of references: such lists exist elsewhere, e.g. Plackett (1981). Cox (1970) offers a good introduction to the subject and has a pleasant blend of theory and application. Breslow and Day (1980) emphasize the applications to retrospective studies. Fleiss (1981) discusses applications to the health sciences generally. Andersen (1973) uses conditional likelihoods in educational testing models. Other books dealing partially or wholly with binary data include Armitage (1971), Ashton (1972), Everitt (1977), Fienberg (1980), Finney (1971), Gokhale and Kullback (1978) and Maxwell (1961). The distinction between the binary response on one hand and explanatory variables on the other has been emphasized here. The books by Bishop *et al.* (1975) and Haberman (1974a) emphasize the log–linear framework of Chapter 6 where the distinction between responses and factors need not be made explicit. Much of the early work on the logistic model is due to Berkson (1944, 1951). Cox (1958a, b) showed how exact inferences could be drawn by the use of conditional arguments providing a connection between Fisher's exact test and McNemar's (1947) test for matched pairs. Recent work in the context of survival data emphasizes the importance of the conditional method in eliminating nuisance parameters. Much of the early work on retrospective studies is due to Mantel, beginning with the important paper by Mantel and Haenszel (1959). Designs for retrospective studies often involve matching of, say, one case with a small number of controls. Here it is essential to use the conditional method of Section 4.4 if misleading conclusions are to be avoided. Adena and Wilson (1982) discuss the application of generalized linear models in epidemiological research.

CHAPTER 5

Polytomous data

5.1 Introduction

If the response of an individual or item in a study is restricted to one of a (usually small) number, say k, of possible values, it is said to be polytomous. The k possible values are referred to as categories: often these are of a qualitative rather than a quantitative nature. Examples include basic blood groups, O, A, B and AB, which are well defined but qualitative, and the I.L.O. scale used for classifying chest X-ray images, 0/0, 0/1, 1/0, ..., 3/3, whose categories are defined rather arbitrarily in terms of 'standard' reproductions. Rating scales used in food testing, which are necessarily subjective, are often of this type. Many variables that arise in the social sciences are, of necessity, not capable of precise measurement and thus fall into the classes discussed here.

To develop a suitable statistical model it is essential to distinguish between the several types of polytomous response or measurement scale. This exercise is usually sufficient to indicate the general class of models likely to be appropriate. Here we are restricting the word 'model' to refer to the structure of the systematic or non-random component: the nature of the random variation is considered separately. The key to simplicity of interpretation usually lies (i) in the correct identification of the response variable and (ii) in using a model suited to the scale of measurement of the response. The choice of response depends on the purpose of the analysis, and a single data set may be used for more than one purpose so that different analyses may use different response variables. In any case correct identification of the measurement scale is likely to simplify the ensuing analysis.

5.1.1 *Measurement scales*

It seems entirely reasonable that the statistical model employed should reflect the nature of the scale of measurement. Thus it is useful, even essential, to classify scales of measurement according to type. Stevens (1951, 1958, 1968) distinguishes nominal, ordinal, interval and ratio scales. Even this list is incomplete; we could, for example, add various types of cardinal scales, e.g. continuous, granular and discrete, or scales where only a subset of the response categories is ordered. Here we distinguish primarily between nominal, ordinal and cardinal scales. Measurements on a cardinal scale are perhaps best handled by the methods of Chapters 3 and 7: we do not consider this type further here.

It is convenient to use the integers $1, \ldots, k$ as labels for the k possible response categories. For nominal responses the integers are simply labels and, as such, the labels could equally well be associated with any permutation of the response categories. For ordinal responses, it is convenient to think of the labels $1, \ldots, k$ as the first k ordinal numbers associated with the categories in their natural order. Of course, the reverse order would do equally well. In neither case does the use of the labels $1, \ldots, k$ imply that the 'spacing' or 'distance' between categories is in any sense regular. We take the view here that the concept of 'distance' or 'spacing' does not apply to scales of measurement that are purely ordinal or purely nominal.

In applications the distinction between ordinal and nominal scales is usually but not always clear. For example, responses relating to food quality – excellent, good, ..., bad, terrible – are clearly ordinal; responses concerning preference of newspaper or television programme would generally be considered to be nominal although for some purposes even these might be considered ordinal. Colours, except where there is a clear connection with the electromagnetic spectrum, would usually be considered to be unordered. However, Pearson (1901), in a study of coat-colour inheritance in thoroughbred horses, takes colour to be an ordinal response variable.

5.1.2 *Models for ordinal scales of measurement*

There are several statistical models specifically designed for ordinal response variables. These can be classified into two types, depending on whether the model is invariant under the grouping of adjacent response categories or not. In many applications, such as taste testing or in the classification of radiographs, the definition of the response categories is entirely arbitrary and often subjective. Thus it appears reasonable to insist that the form of the conclusions should not depend on the particular choice of response categories. In other words if a new category is formed by combining adjacent categories on the old scale, the form of the conclusions should be unaffected. Of course, combining response categories in this way will generally reduce the available information.

These considerations lead to models based not on the individual probabilities for the response categories themselves but rather on the cumulative distribution probabilities γ_j, $j = 1, \dots, k$, where $\gamma_j = \Pi_1 + \dots + \Pi_j$. In particular, linear models using the log-odds scale $\ln[\gamma_j/(1-\gamma_j)]$, the complementary log–log scale $\ln[-\ln(1-\gamma_j)]$ and the log–log scale $\ln(-\ln \gamma_j)$ are found to work well in applications (McCullagh, 1980). Of course the probit scale $\Phi^{-1}(\gamma_j)$ could also be used, but this is very similar to the log-odds scale for much of its range. The simplest models in this class involve parallel regressions on the appropriately chosen scale such as

$$\ln[\gamma_{ij}/(1-\gamma_{ij})] = \theta_j - \boldsymbol{\beta}^{\mathrm{T}}\mathbf{x}_i; \qquad j = 1, \dots, k-1;\ i = 1, \dots, N, \quad (5.1)$$

where the transformation on the left of (5.1) could be replaced by any suitable function mapping the unit interval (0, 1) onto the whole real line. The parameters $\boldsymbol{\theta}$ and $\boldsymbol{\beta}$ are generally unknown.

Note that if (5.1) is true, the interpretation of $\boldsymbol{\beta}$ does not depend on the particular choice of categories used for recording the data. In particular, if several adjacent categories are pooled, the same value of $\boldsymbol{\beta}$ obtains. These comments apply for any suitable link function on the left of (5.1).

We now turn to scales of measurement where the categories are of interest in themselves and where it would not usually make sense to combine adjacent categories. Suitable examples are not common, but the following will illustrate the ideas. Suppose we require to investigate the relative effects of red-clover silage and grass silage on the fertility of lactating cows. We begin with, say, 20 cows on each diet. Most cows become pregnant at the first insemination but a few require a second or third insemination or perhaps more. For each treatment group the observations consist of the number of cows becoming pregnant after one insemination, after two inseminations and so on. We wish to compare the success or pregnancy rates in the two treatment groups.

For the first treatment group, the odds of success or fertility at stage j conditional on failure at all previous stages is $\Pi_{1j}/(1-\gamma_{1j})$: the corresponding odds for the second treatment group being $\Pi_{2j}/(1-\gamma_{2j})$. One simple model for the effect of diet is to suppose that corresponding odds are in constant proportion. In the general notation corresponding to that used in (5.1) with N groups, we obtain

$$\ln[\Pi_{ij}/(1-\gamma_{ij})] = \theta_j - \boldsymbol{\beta}^{\mathrm{T}}\mathbf{x}_i; \qquad j = 1, \ldots, k-1;\ i = 1, \ldots, N. \tag{5.2}$$

Although the same letters are used for the parameters in (5.2) and (5.1) they do not necessarily refer to the same or similar quantities: for example, the θs in (5.1) are required to be monotone increasing while those in (5.2) are unrestricted. In principle the logistic transformation of the conditional probability $\Pi_j/(1-\gamma_{j-1})$ on the left of (5.2) could be replaced by any suitable alternative transformation.

Model (5.2) was proposed by Cox (1972) in the context of survival data in discrete time. In the contingency-table literature it is sometimes referred to as a continuation-ratio model (Fienberg, 1980).

For data arranged in a two-way table with k columns corresponding to the k response categories, Birch (1965) proposed the model

$$\ln \Pi_{ij} = \alpha_i + \beta_j + \delta r_i c_j; \qquad i = 1, \ldots, N;\ j = 1, \ldots, k, \tag{5.3}$$

where r_i and c_j are known scores associated with the rows and columns respectively. Here Π_{ij} is the conditional probability associated with the jth response category in the ith row. It is possible, by letting r depend on the covariate $\mathbf{x}_i$, to extend (5.3) to take some account of covariate information. Model (5.3) has the advantages of mathematical, statistical and computational simplicity but it is difficult to justify along the lines of (5.1) and (5.2). Usually, of course, the scores are unknown; in applications, integers are often used, for example as in Yates (1948), Haberman (1974b) and Goodman (1979). The test statistic associated with H_0: $\delta = 0$ can be expected to have good power against alternatives of most interest (McCullagh, 1980). In recent work Goodman (1981) shows how both row and column scores may be estimated but the justification for such a model from an applied standpoint remains obscure.

The concept of stochastic ordering provides a way of seeing how (5.3) and (5.1) model ordered response categories. For simplicity take

$N = 2$ so that we are concerned with the comparison of two grouped frequency distributions. The corresponding discrete random variables Y_1 and Y_2 are said to be stochastically ordered, $Y_1 \leqslant Y_2$, if $\gamma_{1j} \geqslant \gamma_{2j}$ for each j. Thus stochastic ordering is intimately connected with the cumulative probabilities γ_{ij} or monotone transformations thereof such as occur in (5.1). Similarly, though not so directly, in (5.3), if c_j is increasing in j and $r_2 > r_1$, then $Y_1 \leqslant Y_2$ if $\delta > 0$, and conversely if $\delta < 0$.

It should be emphasized that the choice between the prototype models (5.1), (5.2) and (5.3) can usually be made on external grounds without resorting to goodness-of-fit comparisons. Each prototype allows considerable variety, such as the choice of transformation to linearity and the choice of explanatory variables. Further modifications may be necessary, such as the introduction of non-linear terms on the right of (5.1) (McCullagh, 1980), or (5.3) (Goodman, 1981). Often, of course, a simple informal analysis, graphical or otherwise, using the kind of transformation indicated by (5.1), (5.2) or (5.3), will give all the information required.

5.1.3 *Models for nominal scales of measurement*

For responses that are purely nominal we work with the cell or category probabilities Π_j rather than cumulative or conditional probabilities. In particular this leads to linear models for $\ln \Pi_j$ or $\ln(\Pi_j/\Pi_{j'})$. In other words we write

$$\Pi_j = \exp(\eta_j)/\sum_j \exp(\eta_j); \qquad j = 1, \ldots, k,$$

and examine how the vector $(\eta_1, \ldots, \eta_k)$ depends on the covariates. There is over-parametrization here and we could, for example, set $\hat{\eta}_1 = 0$ or $\Sigma\hat{\eta}_j = 0$. We use the former constraint for convenience.

In many applications the scheme described above requires modification. Suppose, for example, that we wish to compare the preferences of two groups of students for various newspapers. It would usually make sense to conduct such a comparison in two stages: (i) compare the proportions in each group who read newspapers regularly and (ii) compare preferences among regular readers. The second part of this exercise involves conditioning on the observed numbers of regular readers and might well involve the kind of models described above. Of course if the data are not extensive and involve, say, only two groups, a simple informal analysis involving conversion to percentages and the taking of ratios would usually be quite adequate, particularly as no well-founded concise theory is usually available for such examples.

The general log–linear model for nominal response categories can be written as follows. For the ith sample with covariate values $\mathbf{x}_i$, denote the response probabilities by Π_{ij}, $j = 1, \ldots, k$, and write

$$\Pi_{ij} = \exp(\eta_{ij}) \Big/ \sum_j \exp(\eta_{ij}), \tag{5.4}$$

with $\eta_{i1} = 0$ for each i. The systematic part of the model is said to be log–linear if

$$\eta_{ij} = \theta_j + \boldsymbol{\beta}_j^{\mathrm{T}} \mathbf{x}_i; \qquad j = 1, \ldots, k;\ i = 1, \ldots, N \tag{5.5}$$

for unknown θ_j and $\boldsymbol{\beta}_j$. To avoid redundancy we set $\hat{\theta}_1 = 0$ and $\hat{\boldsymbol{\beta}}_1 = \mathbf{0}$. In other words, the net effect of the covariate vector $\mathbf{x}$ relative to the base value $\mathbf{x} = \mathbf{0}$ is to increase the response probabilities by the factors

$$1, \quad \exp(\boldsymbol{\beta}_1^{\mathrm{T}} \mathbf{x}), \quad \ldots, \quad \exp(\boldsymbol{\beta}_k^{\mathrm{T}} \mathbf{x}).$$

Thus the odds in favour of response j over response 1 is increased by the factor $\exp(\boldsymbol{\beta}_j^{\mathrm{T}} \mathbf{x})$. Of course much effort will often go into finding a small subset of suitable terms to include in the vector $\mathbf{x}$.

5.2 Likelihood functions for polytomous data

The details here follow closely those of Sections 4.2 and 4.3 on binary responses. We give the log-likelihood function for multinomial responses. The quasi-likelihood for over-dispersed data is the same as the multinomial log likelihood and the usual asymptotic results hold under second-order assumptions only (see Chapter 8). Again it is important to distinguish clearly between approximations appropriate as a typical (multinomial) index n becomes large and approximations appropriate as the number of (multinomial) observations N becomes large.

5.2.1 *Log likelihood for multinomial responses*

We denote a single multinomial observation by the k-vector $\mathbf{y}_i = (y_{i1}, \ldots, y_{ik})$ with fixed total n. Usually in applications the same

p-vector $\mathbf{x}_i$ of covariates is associated with all k cells of the multinomial observation. This restriction makes sense in almost all applications, although it does not appear to be necessary to the following theory. For log–linear models the restriction is sometimes inconvenient. The multinomial log-likelihood function is most readily expressed in terms of the category probabilities Π_{ij} where $\Sigma_j \Pi_{ij} = 1$. This gives the sum over i of terms of the form

$$l(\mathbf{\Pi}_i; \mathbf{y}_i) = \sum_{j=1}^{k} y_{ij} \log \Pi_{ij}. \tag{5.6}$$

Of course (5.6) could equally well be expressed in terms of the cumulative probabilities $\gamma_{ij} = \Pi_{i1} + \ldots + \Pi_{ij}$. The log likelihood expressed as a function of the unknown parameters $\boldsymbol{\theta}$ and $\boldsymbol{\beta}$, say, can be found by substituting the appropriate systematic model (5.1), (5.2) or (5.3) into the log likelihood $l(\mathbf{\Pi}; \mathbf{y}) = \Sigma_i l(\mathbf{\Pi}; \mathbf{y}_i)$. We do not do so explicitly.

In the special cases (5.3) (with r_i and c_j not both unknown) and (5.5), where the model is log–linear, sufficiency induces a simplification of the log likelihood. Special methods such as iterative proportional fitting (Deming and Stephan, 1940; Fienberg 1970a; Darroch and Ratcliff, 1972) can then be used for fitting and parameter estimation. However, because these methods are readily available in standard computing packages and because we do not wish to emphasize computational over statistical issues, we restrict attention to methods of estimation that are more widely applicable. In effect this means iterative weighted least squares. In certain special cases, particularly if the number of parameters is large, other methods of computation may well prove more efficient and computationally more stable.

The fitting method is most easily described via quasi-likelihood functions which we now introduce in the context of polytomous data.

5.2.2 *Quasi-likelihood functions for over-dispersed data*

In Chapter 8 we give a justification for the use of the multinomial log-likelihood function even in cases where the multinomial distribution does not strictly apply. This justification is based on an assumption that the total of the observations over the k response categories is fixed and that

$$\mathrm{E}(\mathbf{Y}) = n\mathbf{\Pi} \qquad \text{and} \qquad \mathrm{cov}(\mathbf{Y}) = n\sigma^2\mathbf{V}, \tag{5.7}$$

where $\mathbf{\Pi}$ is a vector of probabilities and $n\mathbf{V}$ is the covariance matrix of a multinomial observation with parameter vector $\mathbf{\Pi}$. The dispersion parameter σ^2 often exceeds unity, representing over-dispersion relative to the multinomial model. Quasi-likelihood functions can in fact be defined for arbitrary positive semi-definite covariance matrices but the explicit functional form is rarely easy to find.

For ordinal response variables it may sometimes be more convenient to deal with the cumulative observations, $z_j = y_1 + \ldots + y_j$. In matrix notation we may write $\mathbf{z} = \mathbf{Ly}$ where $\mathbf{L}$ is a lower-triangular matrix. The mean vector and covariance matrix of $\mathbf{Z}$ are $n\gamma = n\mathbf{L\Pi}$ and $\sigma^2\mathbf{LVL}^T$. The quasi-likelihood function is invariant under invertible linear transformations and so it makes no difference whether we use $\mathbf{Z}$ or $\mathbf{Y}$ as our basic observation.

The log likelihood or quasi-likelihood for a single response vector $\mathbf{y}$ having k categories is

$$l(\mathbf{\Pi}; \mathbf{y}) = \Sigma y_j \log \Pi_j, \tag{5.8}$$

it being understood that $\Sigma \Pi_j = 1$. The derivative of l with respect to $\mathbf{\Pi}$ satisfies

$$\frac{\partial l}{\partial \mathbf{\Pi}} = \mathbf{V}^{-}(\mathbf{y} - n\mathbf{\Pi}), \tag{5.9}$$

the defining equation for a quasi-likelihood function. Suppose now that the systematic part of the model can be written $\mathbf{\Pi} = \mathbf{\Pi}(\boldsymbol{\beta})$. The estimating equations for $\boldsymbol{\beta}$ become

$$\mathbf{D}^T\mathbf{W}(n^{-1}\mathbf{y} - \mathbf{\Pi}) = 0, \tag{5.10}$$

where $\mathbf{W} = n\mathbf{V}^{-}$ is a matrix of weights and $\mathbf{D} = \mathrm{d}\mathbf{\Pi}/\mathrm{d}\boldsymbol{\beta}$ is a derivative matrix of order $k \times p$. If several independent multinomial observations are available, the estimating equations involve a sum of terms each having the form (5.10). The information matrix for the parameters is the sum over the multinomial observations of terms having the form

$$\sigma^{-2}(\mathbf{D}^T\mathbf{WD}). \tag{5.11}$$

The variances and covariances of the parameter estimates include the dispersion parameter σ^2 which should be set equal to unity if the data have the multinomial distribution. Otherwise σ^2 may need to be estimated from the data. Appropriate methods are now discussed.

5.2.3 *Asymptotic theory*

The quasi-likelihood coincides with the multinomial log likelihood (5.6). It follows therefore that the validity of the usual asymptotic results based on $l(\mathbf{\Pi}; \mathbf{y})$ depends only on second-moment assumptions. To the usual order of approximation in which, for example, the parameter estimates are taken as Normally distributed, higher moments of $\mathbf{Y}$ are irrelevant. Another aspect of the same thing is that faulty or misrecorded observations may exert an undue influence on the parameters. Consequently an examination of the residuals, either formally or informally, is essential. We illustrate this primarily via the examples in Section 5.3.

As with binary data it is essential to distinguish between asymptotic results that apply only to grouped data and results that apply to ungrouped data. Thus the distribution of the likelihood ratio and Pearson's goodness-of-fit statistics can be taken as χ^2 only if each of the (multinomial) indices is large. On the other hand, the asymptotic covariance matrix (5.11) for the parameter estimates applies with either asymptote.

To estimate σ^2 we use the method described in Chapter 4 which is based on the Pearson statistic

$$X^2 = \sum_{ij} (y_{ij} - \hat{\mu}_{ij})^2 / \hat{\mu}_{ij}, \qquad (5.12)$$

the sum extending over all Nk cells. Let p be the total number of parameters including, for example, all the elements of $\boldsymbol{\theta}$ and $\boldsymbol{\beta}$ in (5.1) or (5.2). Then a consistent estimator for σ^2 is given by

$$\tilde{\sigma}^2 = X^2/(Nk - N - p). \qquad (5.13)$$

Here, by analogy with (4.14), the effective degrees of freedom for $\tilde{\sigma}^2$ are $[N(k-1)-p]/(1+\frac{1}{2}\bar{\gamma}_2)$ where $\bar{\gamma}_2$ is the average standardized kurtosis of the observations and is $O(1/n)$. When n is moderately large this refinement can be neglected.

5.3 Examples

5.3.1 *A tasting experiment*

The following data (kindly provided by Dr Graeme Newell) are derived from an experiment concerned with the effect on taste of various cheese additives. The nine response categories range from strong dislike (1) to excellent taste (9). Four additives labelled A, B, C and D were tested (Table 5.1).

Here the effects are so great that the qualitative ordering (D, A, C, B) can be deduced purely from visual inspection. Nevertheless it is of some interest to check whether the models described earlier are capable of describing these observed differences and to evaluate the significance of the differences observed.

Table 5.1 *Response frequencies in a cheese tasting experiment*

	Response category									
Cheese	1*	2	3	4	5	6	7	8	9†	*Total*
A	0	0	1	7	8	8	19	8	1	52
B	6	9	12	11	7	6	1	0	0	52
C	1	1	6	8	23	7	5	1	0	52
D	0	0	0	1	3	7	14	16	11	52
Total	7	10	19	27	41	28	39	25	12	208

* 1 = strong dislike.
† 9 = excellent taste.
Data courtesy of Dr G. Newell, Hawkesbury Agricultural College.

The nature of the response is such that a model of the form (5.1) is most appropriate. We try first the logistic transformation although, of course, an alternative transformation might give a better fit. In fact the probit transformation is better than the logit but not greatly so. The appropriate version of (5.1) involves parameters $\theta_1, \ldots, \theta_8$, the βs being the coefficients of four indicator variables, one for each treatment. One level of the treatment factor is redundant and can be set to zero. Thus (5.1) can be written

$$\log[\gamma_{ij}/(1-\gamma_{ij})] = \theta_j - \beta_i \tag{5.14}$$

with, say, $\hat{\beta}_1 = 0$ or $\Sigma_i \hat{\beta}_i = 0$.

As it happens, we have set $\hat{\beta}_4 = 0$ so that our estimates are logistic contrasts with treatment D. The estimates, standard errors and correlation matrix of the $\hat{\beta}$s are given below.

		SE	Correlations		
A	$\hat{\beta}_1 = -1.613$	0.378	1.0		
B	$\hat{\beta}_2 = -4.965$	0.474	0.525	1.0	
C	$\hat{\beta}_3 = -3.323$	0.425	0.574	0.659	1.0
D	$\hat{\beta}_4 = 0.0$		–	–	–

Positive values of β represent a tendency towards the higher-numbered categories relative to D: negative values indicate the reverse. These values confirm the ordering (D, A, C, B). The standard errors are based on the assumption that $\sigma^2 = 1$.

The deviance for these data is reduced from 168.8 on 24 degrees of freedom under the model of no preference ($\boldsymbol{\beta} = 0$) to 20.31 on 21 degrees of freedom under (5.14). Because of the small numbers in the extreme cells the chi-squared approximation for the deviance is not very good here. Residual analysis is awkward partly for the same reason and partly because row sums are fixed. Using the crude standardization $(y_{ij} - \hat{y}_{ij})/[n_i \hat{\Pi}_{ij}(1 - \hat{\Pi}_{ij})]^{\frac{1}{2}}$ we find two cell residuals exceeding the value 2.0. The values are 2.23 and 2.30 corresponding to cells (1, 4) and (2, 6) with fitted values 3.16 and 2.47 respectively. However, if residual calculations were based on the cumulative totals $z_{ij} = y_{i1} + \ldots + y_{ij}$, arguably a more appropriate procedure here, the apparently large discrepancies would disappear. At least residuals based on z_{ij} have the strong conceptual advantage that only $k-1$ of them are defined for each multinomial observation. Correlation among the residuals is a problem regardless of definition but the problem seems more acute for residuals based on z_{ij}. In any case we do not regard these extreme residuals as sufficiently large to refute the model which is, at best, an approximation to reality.

So far we have assumed $\sigma^2 = 1$ without justification. However, the estimate for σ^2 from (5.13) is 20.9/21 or almost exactly unity. Here $p = 11$ is the total number of parameters including the θs.

A more serious problem that we have so far ignored is that observations corresponding to different treatments are not independent. The same 52 panellists are involved in all four tests. This is likely to induce some positive correlation ρ between the ratings for the different treatments. Variances of contrasts would then be reduced by a factor $1 - \rho^2$ relative to independent measurements. Inferences based on supposing that $\rho = 0$ are therefore conservative. In other words the general qualitative and quantitative conclusions remain valid with the computed variances being regarded as approximate upper limits.

Finally we examine the effect on $\hat{\boldsymbol{\beta}}$ of reducing the number of response categories. Various combinations are possible: here we combine categories 1, ..., 4 and 7, 8, 9, thus reducing the original nine categories to four. This arrangement makes all cell counts positive. The new estimates for $\boldsymbol{\beta}$ are $(-1.34, -4.57, -3.07, 0)$ corresponding to an average reduction of about 0.7 standard errors compared with the previous analysis. Reduction of the number of categories does not always have this effect. Estimated variances are increased by an average of about 19%. Correlations are virtually unaffected.

The available evidence suggests that, when the data are sparse, the estimate of $\hat{\beta}_j$ may be too large in magnitude. Grouping of the tail categories has the effect of reducing this bias. A very small-scale simulation experiment based on 25 repetitions using the values of $\hat{\boldsymbol{\theta}}$ and $\hat{\boldsymbol{\beta}}$ obtained from the data in Table 5.1 and the same row totals indicates the following:

1. The bias in the estimates $\hat{\beta}_j$ is no more than 5%.
2. The deviance or likelihood-ratio goodness-of-fit statistic is approximately distributed as χ^2_{21}: at least the first two moments do not differ appreciably from those of this reference distribution.
3 The standard errors obtained from the diagonal elements of (5.11) are, if anything, a little too large – by about 10%.

Because of the small scale of the simulation, these conclusions are expressed with a certain degree of caution. Nonetheless the conclusions are positive and they show that, even with data as sparse as those in Table 5.1 and where the number of parameters (11) is moderately large in comparison to the number of observations (32), the usual asymptotic results are quite reliable at least for the parameters of primary interest.

5.3.2 *Pneumoconiosis among coalminers*

The following example illustrates the use of a quantitative covariate in an ordinal regression model. For comparative purposes we apply both (5.1) and (5.2). Difficulties associated with residual plots are also illustrated.

The data are taken from Ashford (1959) and concern the degree of pneumoconiosis in coalface workers. Here, the severity of the

Table 5.2 *Period of exposure and prevalence of pneumoconiosis amongst a group of coalminers*

	Number of men		
Period spent (yr)	*Category 1: normal*	*Category 2*	*Category 3: severe pneumoconiosis*
5.8	98	0	0
15.0	51	2	1
21.5	34	6	3
27.5	35	5	8
33.5	32	10	9
39.5	23	7	8
46.0	12	6	10
51.5	4	2	5

disease is measured radiologically and is, of necessity, qualitative. The observations, as set out in Table 5.2, are grouped according to the length of time, t, that individuals have spent working in the mine. Originally there were four response categories, ranging from normal to severe pneumoconiosis, but here the third and fourth categories have been combined.

A preliminary plot of the transformed variables

$$\ln\left(\frac{y_{i1}+\frac{1}{2}}{n_i-y_{i1}+\frac{1}{2}}\right) \quad \text{and} \quad \ln\left(\frac{y_{i1}+y_{i2}+\frac{1}{2}}{n_i-y_{i1}-y_{i2}+\frac{1}{2}}\right) \tag{5.15}$$

against t_i reveals approximately parallel but non-linear relationships. Further investigation shows that the transformed variables (5.15) are approximately linear in $\ln t_i$. We are thus led to consider the model

$$\ln[\gamma_{ij}/(1-\gamma_{ij})] = \theta_j - \beta \ln t_i; \qquad j = 1, 2;\ i = 1, \ldots, 8. \tag{5.16}$$

We might expect that the non-linearity of (5.16) in t could have been detected by an appropriate analysis of the residuals after fitting the model linear in t. This is indeed so but some care is required. When the 24 cell residuals, appropriately standardized, are plotted against t_i no strong curvilinear pattern is discernible. On the other hand, a plot against t_i of the cumulative residuals, $y_{i1}-\hat{y}_{i1}$ and $y_{i1}+y_{i2}-\hat{y}_{i1}-\hat{y}_{i2}$, appropriately standardized, clearly reveals the non-linearity. When $k = 3$ this is equivalent to ignoring the residuals associated with category 2 and changing the sign of the category-3

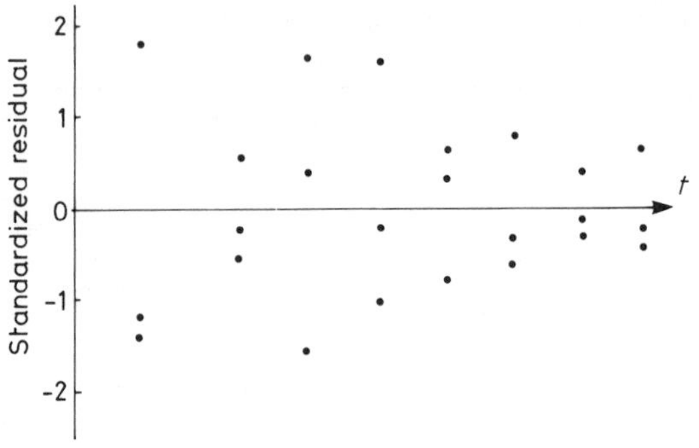

Fig. 5.1. *Plot of individual residuals against t for the pneumoconiosis data.*

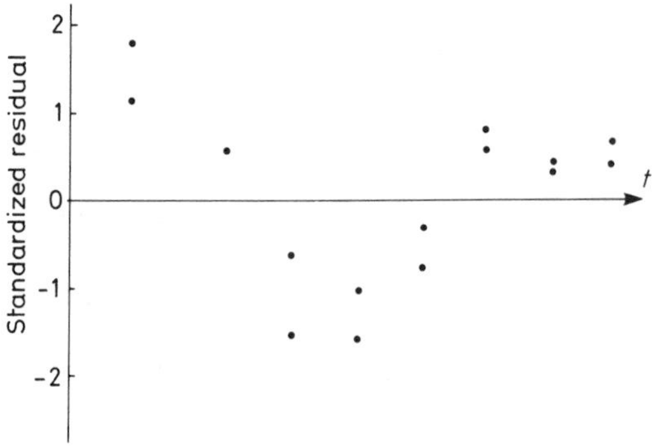

Fig. 5.2. *Plot of cumulative residuals against t for the pneumoconiosis data.*

residuals. The two plots are displayed in Figs 5.1 and 5.2 respectively. The simplified standardization used here takes no account of the errors involved in using estimated values of the parameters.

The analysis using (5.16) gives a value of $\hat{\beta}$ of 2.594 with standard error 0.38, while the values of $\hat{\theta}_1$ and $\hat{\theta}_2$ are 9.667 and 10.572 respectively. No pattern is discernible among the residuals and the fit is good. The conclusions, therefore, are that for a miner with, say, five years of exposure the odds of having pneumoconiosis are one in $\exp(9.667-2.594 \ln 5)$, i.e. one in 243. Doubling the exposure increases the risk by a factor of $2^{2.594} = 6.0$, so that after 10 years the risk rises to just under one in 40, and after 20 years to just over one in seven. For severe pneumoconiosis the five-year risk is one in $\exp(10.572+2.594 \ln 5)$, i.e. one in 600. This risk increases at the same rate as before so that after 10 years exposure the estimated risk is one in 100 and after 20 years is one in 17.

Ashford's analysis of these data proceeds along the same lines as that given here except that he uses the probit function in place of the logit. His conclusions give similar fitted values but his parameter estimates are different, partly because of his use of the probit function and partly because of numerical inaccuracies.

We now proceed to illustrate the use of the alternative model (5.2) in the context of the pneumoconiosis data. The odds in favour of category 1 are $\exp(\theta_1 - \beta x_i)$ so that the odds or risk of having the disease is $\exp(-\theta_1 + \beta x_i)$. Here x_i is a general measure of exposure – in this case t_i or $\ln t_i$. Thus the risk for the disease increases by the factor e^β per unit increase in x. Among those who have the disease, the risk, or odds of having severe symptoms, is $\exp(-\theta_2 + \beta x_i)$, so that again the risk increases by the factor e^β per unit increase in x. There is clearly the possibility that a different β might be involved in the second expression.

Because of the special structure of the model (5.2) each trinomial observation can be broken into two binomial components. The first component specifies the number of diseased individuals as proportion of the total number at risk, while the second component gives the number of severely diseased as a proportion of those with the disease. Thus

$$\begin{aligned} y_{11} &= 0/98, & y_{12} &= 0/0, \\ y_{21} &= 3/54, & y_{22} &= 1/3, \\ y_{31} &= 9/43, & y_{32} &= 3/9, \end{aligned}$$

and so on.

These binomial observations y_{ij}/n_{ij} can be regarded as independent observations with probabilities Π_{ij} satisfying

$$\ln[\Pi_{ij}/(1 - \Pi_{ij})] = -\theta_j + \beta x_i; \qquad j = 1, 2;\ i = 1, \ldots, 8, \quad (5.17)$$

bearing in mind that the relationships may not be parallel. In other words it may be necessary to write β_j instead of β in (5.17).

As it happens, the covariate $\ln t_i$ is strongly preferred to t_i for these data; there is inconclusive evidence on whether $\beta_1 \neq \beta_2$. As measured by the deviance or likelihood-ratio statistic, the fit of (5.17) is comparable to that of (5.16). The goodness-of-fit statistics are 5.1 on 13 d.f. for (5.16) and 7.6 on 12 d.f. for (5.17). One degree of freedom is lost because y_{12} is degenerate or non-random when (5.17) is used. For model (5.17) the residuals give some indication of a faint pattern;

for this reason the former model (5.16) might be preferred. In any case the estimate of β is 2.321 with approximate standard error 0.33, these values being similar to those obtained earlier despite the slight difference of interpretation. Thus we are led to the estimate of $2^{2.321} = 5.0$ as the increase in risk associated with doubling the exposure time.

To summarize we can say that (5.1) and (5.2), or equivalently (5.16) and (5.17), are different ways of describing the risk associated with increasing exposure. The conclusions from either analysis support the claim that doubling the exposure increases the risk by an estimated factor of between 5 and 6. Approximate 95% confidence limits for this factor are (3.2, 10.2). It would be of interest to know (i) whether the risk would continue to increase if exposure were to cease, and (ii) whether the risk would increase more slowly if dust levels were reduced. The data given here do not allow us to investigate these questions; indeed as the data stand, such effects would be likely to be confounded with age.

* 5.4 Conditional and other likelihoods for the elimination of nuisance parameters

All the models for multinomial responses so far discussed involve some parameters of little or no interest in answering the main question. We refer to such quantities as nuisance or incidental parameters. In a single $2 \times k$ table no particular difficulty arises with these nuisance parameters provided that they can be consistently estimated. A major difficulty arises when information from several sparse tables is to be combined. Here the number of nuisance parameters can increase dramatically and the usual asymptotic results of Section 5.2 cease to apply and can even be misleading.

In this section we discuss the construction of conditional likelihoods defined so that the nuisance parameters are eliminated. The appropriate conditional likelihood depends very much on the form of model thought appropriate. The discussion here refers to a single table only. When the data involve several such tables a modified log-likelihood function not involving the nuisance parameters can be formed by adding the contributions from each table separately.

5.4.1 *Proportional-odds models*

We consider the proportional-odds model (5.1) for ordinal measurement scales, using, for simplicity, the comparison of two multinomial observations with totals n_1 and n_2 and response category totals $m_1, \ldots, m_k$. Let $n = n_1 + n_2 = m_1 + \ldots + m_k$. In terms of the cumulative probabilities γ_{1j} and γ_{2j}, the model can be written

$$\begin{aligned} \log[\gamma_{1j}/(1-\gamma_{1j})] &= \theta_j; \\ \log[\gamma_{2j}/(1-\gamma_{2j})] &= \theta_j - \Delta; \end{aligned} \qquad j = 1, \ldots, k-1. \tag{5.18}$$

Here Δ is the parameter of interest and we wish to eliminate the $k-1$ nuisance parameters $\theta_1, \ldots, \theta_{k-1}$. The observations associated with the first sample can be denoted by $y_1, \ldots, y_k$ with $\Sigma y_j = n_1$ and, because (5.18) is expressed in terms of cumulative probabilities, it is convenient to write $z_j = y_1 + \ldots + y_j$ and $M_j = m_1 + \ldots + m_j$. The random variable Z_j is binomially distributed with index n_1 and parameter γ_{1j}. Now, for any fixed j the distribution of Z_j given M_j is hypergeometric, depends only on Δ, and can be written

$$\text{pr}(Z_j = z_j | M_j) = \binom{n_1}{z_j}\binom{n_2}{M_j - z_j}\psi^{z_j} \Big/ \sum_s \binom{n_1}{s}\binom{n_2}{M_j - s}\psi^s, \tag{5.19}$$

where $\psi = e^{\Delta}$ and the sum in the denominator of (5.19) extends over all non-negative integers consistent with the marginal totals M_j, n_1, n_2. Thus by making probability calculations using the conditional distribution (5.19), nuisance parameters can be eliminated. Unfortunately this can be done in $(k-1)$ ways, each of which involves some considerable loss of information. It would seem that these $(k-1)$ sources of information ought to be combined.

The following method seems plausible and works well in most applications. Let μ_j and V_j, both functions of Δ, be the conditional mean and variance of Z_j computed from (5.19). We will later use the fact that $V_j = d\mu_j/d\Delta$. Note that because Z_j and $Z_{j'}$ $(j \neq j')$ are random variables on different sample spaces, it is not generally possible to talk of the covariance between Z_j and $Z_{j'}$. However, in the special case $\psi = 1$ we can condition simultaneously on $M_1, \ldots, M_k$. The exact correlation matrix $\mathbf{C}$ of $Z_1, \ldots, Z_{k-1}$ has elements c_{ij} given by

$$c_{ij} = \left(\frac{g_i(1-g_j)}{g_j(1-g_i)}\right)^{\frac{1}{2}}; \qquad i \leqslant j, \tag{5.20}$$

where $g_j = M_j/n$. It seems plausible that (5.20) may hold in some approximate sense for $\psi \neq 1$. The implications of this are now explored.

Probably the simplest way to estimate Δ is to choose the value $\hat{\Delta}_c$ such that the sum of the zs is equal to the sum of their conditional expectations. In other words to solve

$$\Sigma z_j = \Sigma \mu_j(\hat{\Delta}_c). \tag{5.21}$$

The estimate given by (5.21) is always consistent, is easy to compute, but is slightly inefficient. Let $\mathbf{V}$ be the approximate covariance matrix of the Zs with diagonal elements V_j and off-diagonal elements $c_{ij}(V_i V_j)^{\frac{1}{2}}$. The approximate asymptotic variance of $\hat{\Delta}_c$ is given by

$$\text{var}(\hat{\Delta}_c) = (\mathbf{1}^T\mathbf{V}\mathbf{1})/(\Sigma V_j)^2$$

where $\mathbf{1}$ is the unit vector.

The following simple artificial example illustrates the calculations involved. Suppose $k = 3$ and that we have the following numbers:

	Response category			
	I	II	III	Total
Sample 1	1	3	6	10
Sample 2	3	4	3	10
Total	4	7	9	20

It is convenient to separate the numbers into two tables as follows:

	I	II and III	Total
	1	9	10
	3	7	10
Total	4	16	20

I and II	III	Total
4	6	10
7	3	10
11	9	20

Here we have $g_1 = 4/20 = 0.2$, $g_2 = 11/20 = 0.55$, giving the correlation $r_{12} = r_{21} = 0.4523$. Starting with $\Delta = 0$, $\psi = 1$ we find $\mu_1 = 2.0$, $\mu_2 = 5.5$,

$$V_1 = \frac{4 \times 16 \times 10 \times 10}{20 \times 20 \times 19} = 0.8421,$$

$$V_2 = \frac{11 \times 9 \times 10 \times 10}{20 \times 20 \times 19} = 1.3026,$$

giving $\Sigma(z_j - \mu_j) = 2.5$. The next estimate of Δ is

$$\hat{\Delta}_1 = \Sigma(z_j - \mu_j)/\Sigma V_j = -1.1657.$$

On the next cycle we find $\mu_1 = 1.798$, $\mu_2 = 4.0243$, $V_1 = 0.6894$, $V_2 = 1.1948$, giving $\hat{\Delta}_2 = -1.2210$. Finally we obtain $\hat{\Delta}_c = -1.2214$, the same value that would be obtained if the pair of tables given above were independent. However, the variance of $\hat{\Delta}_c$ reflects the non-independence. We find, using $V_1 = 0.6760$ and $V_2 = 1.1848$, that $\text{var}(\hat{\Delta}_c) = 0.7712$, as opposed to $(\Sigma V_j)^{-1} = 0.5374$ if the tables were independent.

The unconditional estimate obtained from these numbers is $\hat{\Delta}_u = -1.280$ with approximate variance 0.7815. The discrepancy of 7% of a standard error is unlikely to cause serious ambiguities, at least when only a single table is involved and the counts are of a magnitude at least as great as those used here.

When several $2 \times k$ tables are involved, with different nuisance parameters but common value Δ in each table, the elimination of the nuisance parameters becomes essential. The estimating equation is found by adding (5.21) over each of the tables and can be written

$$\Sigma\Sigma z_{ij} = \Sigma\Sigma\mu_{ij}(\hat{\Delta}_c)$$

where z_{ij} is the cumulative total for the ith table and μ_{ij} is its conditional expectation. The asymptotic variance of $\hat{\Delta}_c$ is

$$\text{var}(\hat{\Delta}_c) = \sum_i (\mathbf{1}^T\mathbf{V}^{(i)}\mathbf{1}) \Big/ \left(\sum_i \sum_j V_j^{(i)}\right)^2 = \mathbf{1}^T\mathbf{V}^{(\cdot)}\mathbf{1} \Big/ \left(\sum_j V_j^{(\cdot)}\right)^2$$

where $\mathbf{V}^{(i)}$ is approximate covariance matrix of the Zs for the ith table and $\mathbf{V}^{(\cdot)}$ is the sum of these approximate covariance matrices.

In the matched-pairs problem, where conditioning is essential if misleading conclusions are to be avoided, and where $\hat{\Delta}_u$ tends in probability to 2Δ, the conditional estimator $\hat{\Delta}_c$ can be found in the following way. For each j let R_j be the number of pairs for which the first observation exceeds j and the second is in category j or below. Conversely, $N_j - R_j$ is the number of pairs for which the first observation is category j or below and second category $j+1$ or above. Similarly, for $i \leqslant j$, N_{ij} is the number of pairs for which the response is $(\leqslant i, > j)$ or $(> j, \leqslant i)$ in an obvious notation. From (5.18) and using the results of Cox (1958b) we have that R_j given N_j is binomially distributed with parameter $e^{\Delta}/(1+e^{\Delta})$, giving $k-1$ estimates $\hat{\Delta}_j = \ln[R_j/(N_j - R_j)]$. The combined estimate based on (5.21) is $\hat{\Delta}_c = \log[R_./(N_. - R_.)]$. Write $\hat{p} = R_./N_.$. The approximate covariance matrix $\mathbf{V}^{(\cdot)}$ has elements given by $N_{ij}\hat{p}(1-\hat{p})$, so that the asymptotic variance of $\hat{\Delta}_c$ is

$$[\hat{p}(1-\hat{p})]^{-1}\left(\sum_{ij} N_{ij}\right)\Big/ N_{.}^{2}.$$

A different but slightly more efficient method of combination was given by McCullagh (1977).

Table 5.3 *Unaided distance vision of 7477 women aged 30–39*

	Grade of left eye			
Grade of right eye	*Highest*	*Second*	*Third*	*Lowest*
Highest	1520	266	124	66
Second	234	1512	432	78
Third	117	362	1772	205
Lowest	36	82	179	492

To take a very simple example, consider the distance–vision data of Stuart (1953) given in Table 5.3. These data have been analysed by a number of authors including Plackett (1981) and Bishop *et al.* (1975). The data involve measurements of vision quality in the left and right eyes and clearly involve matched observations. By making a dichotomy at $j = 1$, $j = 2$ and $j = 3$, three 2×2 tables can be constructed. These are:

$$\begin{matrix} 1520 & 456 \\ 387 & 5114, \end{matrix} \qquad \begin{matrix} 3532 & 700 \\ 597 & 2648 \end{matrix} \qquad \text{and} \qquad \begin{matrix} 6339 & 349 \\ 297 & 492. \end{matrix}$$

The three conditional estimates ignore the diagonal observations in each table to give $\hat{\Delta}_1 = \ln(456/387) = 0.164$, $\hat{\Delta}_2 = \ln(700/597) = 0.159$ and $\hat{\Delta}_3 = \ln(349/297) = 0.161$ which are remarkably close, perhaps too close.

The estimate $\hat{\Delta}_c$ obtained by the conditional method described above is $\hat{\Delta} = \ln(1505/1281) = 0.161$. The approximate standard error of $\hat{\Delta}_c$ is 0.0426.

5.4.2 *Proportional-hazards or continuation-ratio models*

A similar kind of conditional analysis can be made for models of the form (5.2). Again we consider the case of two independent multinomial observations. The model can be written

$$\log[\Pi_{1j}/(1-\gamma_{1j})] = \theta_j, \qquad (5.22)$$
$$\log[\Pi_{2j}/(1-\gamma_{2j})] = \theta_j - \Delta.$$

The parameters involved here are not to be confused with similarly denoted quantities in (5.18). Model (5.22) is most often used in the context of survival analysis and the terminology used here is derived from that application. Other aspects of survival data are discussed in Chapter 9.

We imagine the categories as failure times in increasing order. Thus m_j is the total number of failures at time j, $j = 1, 2, \ldots$. Those individuals available for failure at time j constitute the risk set at time j and comprise s_{1j} individuals from sample 1 and s_{2j} individuals from sample 2. Thus s_{ij} is the total number of individuals from sample i failing at time j or later. Let y_j be the number of failures from sample 1 at time j; then the distribution of the random variable Y_j given m_j, s_{1j} and s_{2j} is hypergeometric and depends only on Δ, not on $\boldsymbol{\theta}$. Thus

$$\mathrm{pr}(Y_j = y_j | m_j, s_{1j}, s_{2j}) = \binom{s_{1j}}{y_j}\binom{s_{2j}}{m_j - y_j}\psi^{y_j} \Big/ \sum_r \binom{s_{1j}}{r}\binom{s_{2j}}{m_j - r}\psi^r, \qquad (5.23)$$

where $\psi = e^{\Delta}$.

Again, as in the previous section, there is the conceptual difficulty that the conditioning events in the $k-1$ conditional distributions (5.22) vary. In fact, if we take the intersection of the conditioning events, little randomness remains. Cox (1972, 1975) justifies taking the product of the $(k-1)$ quantities in (5.23) and refers to the resulting quantity as a partial likelihood. In the present context this procedure is equivalent to forming $(k-1)$ 2×2 tables, the jth table being formed by discarding columns $1, \ldots, j-1$ and combining columns $j+1, \ldots, k$. The $(k-1)$ tables are then regarded as independent with common odds ratio ψ. The methods of Section 4.4.1 can now be used.

5.4.3 *Log–linear models*

We now consider ways of eliminating nuisance parameters in log–linear models of the form (5.3) or (5.5). When the data comprise two independent multinomial observations, the parameters of interest involve the $(k-1)$ odds ratios

$$\psi_j = \Pi_{1j}\Pi_{2k}/(\Pi_{2j}\Pi_{1k}).$$

The sufficient statistic for the nuisance parameters is the vector of category totals $\mathbf{m} = (m_1, \ldots, m_k)$. The required conditional distribution of the data given $\mathbf{m}$ is

$$\mathrm{pr}(\mathbf{y}|\mathbf{m}) = \binom{n_1}{\mathbf{y}}\binom{n_2}{\mathbf{m}-\mathbf{y}}\psi_1^{y_1}\ldots\psi_{k-1}^{y_{k-1}} \Big/ \sum_{\mathbf{j}} \binom{n_1}{\mathbf{j}}\binom{n_2}{\mathbf{m}-\mathbf{j}}\psi_1^{j_1}\ldots\psi_{k-1}^{j_{k-1}}, \tag{5.24}$$

where

$$\binom{n}{\mathbf{y}} = n!/(y_1!\ldots y_k!)$$

and the sum in the denominator extends over all non-negative values compatible with the given marginal totals. Note that under (5.3) the parameters ψ_j satisfy the log–linear model

$$\ln \psi_j = \gamma(c_j - c_k), \tag{5.25}$$

where $\{c_j\}$ are known category scores. In fact (5.24) is an exponential family model and (5.25) constitutes a linear model in the family. Fitting and testing are straightforward in principle, though unfortunately the moments of (5.24) are difficult to compute except in the special or 'null' case $\psi_1 = \ldots = \psi_{k-1} = 1$. Apart from this null case the conditional distributions associated with (5.24) are multivariate hypergeometric, like (5.24), but the marginal distributions are not. In the null case the moments of $\mathbf{Y}$ are given by

$$\begin{aligned} \mathrm{E}(Y_j) &= m_j n_1/n, \\ V_{jj} = \mathrm{var}(Y_j) &= m_j(n-m_j)n_1 n_2/[n^2(n-1)], \\ V_{ij} = \mathrm{cov}(Y_i\, Y_j) &= -m_i m_j n_1 n_2/[n^2(n-1)]. \end{aligned} \tag{5.26}$$

In many applications where the response categories are ordered, no scores are readily available; simple integer scores are often used but this choice does not seem entirely satisfactory. However, some progress can be made. We can expand the proportional-odds model (5.18) in a Taylor series about $\Delta = 0$ to get

$$\log \psi_j = -\Delta(c_j - c_k) + \Delta^2(d_j - d_k)/2 + \ldots \tag{5.27}$$

with $c_j = \gamma_j + \gamma_{j-1}$ and $d_j = \gamma_j(1-\gamma_j) + \gamma_{j-1}(1-\gamma_{j-1})$. Here M_j/n is a reasonable estimator of γ_j, suggesting the use of scores proportional to the average category rank. There is a close connection here with the use of ridit scores (Bross, 1958). In (5.23) the totals M_j and hence the scores c_j are regarded as fixed. No allowance is therefore required for errors of estimation. Some loss of power seems inevitable, particularly if the category totals m_j are small.

As an example of how this conditional method might be used, we take the numbers used in Section 5.4.1 (a single 2×3 table) with scores $c = 0.20$, $c = 0.75$, $c = 1.55$ as suggested by the preceding analysis. We suppose that the appropriate model is (5.25) or (5.27) with the quadratic term eliminated. The likelihood equation for Δ is

$$\frac{\partial l}{\partial \Delta} = -\sum_{j=1}^{k-1} (c_j - c_k)(y_j - \mu_j) = 0,$$

and the iterative process for estimation is given by

$$\hat{\Delta}_{j+1} = \hat{\Delta}_j + A^{-1} \frac{\partial l}{\partial \Delta}$$

with

$$A = \sum_i \sum_j (c_j - c_k) V_{ij} (c_j - c_k).$$

Starting with $\Delta = 0$ we find $\partial l / \partial \Delta = -1.75$ and $A = 1.5059$ giving $\hat{\Delta}_1 = -1.1621$. In principle we could repeat the cycle until convergence occurs. Note that the values obtained on the first cycle here are very similar to those obtained in Section 5.4.1. If arbitrary scores were used this similarity would disappear.

When information is available from several strata independently, the appropriate conditional likelihood is the product of terms (5.3), one for each stratum. The resulting function is unpleasant in the extreme, particularly from a computational viewpoint. However, some progress can be made in certain special cases, particularly those of matched pairs or matched triplets.

For matched pairs, $n_1 = n_2 = 1$, and conditioning on $\mathbf{m}$ simply specifies for each pair the two categories i and j that were observed. If the responses happen to be identical so that only one category is observed, the required conditional distribution is degenerate. Otherwise the conditional probability that it was the first member of the pair who responded in category i is $\psi_i/(\psi_i + \psi_j)$. The required conditional likelihood involves a product of such factors. Partly for convenience, data from a matched-pairs design are usually recorded as a square table of counts t_{ij} where t_{ij} is the total number of ordered responses (i, j). The conditional likelihood takes $t_{ij} + t_{ji}$ as fixed so that the conditional distribution of the random variable T_{ij} is binomial with index $t_{ij}+t_{ji}$ and parameter $\psi_i/(\psi_i+\psi_j)$. Thus the conditional likelihood involves a product of $k(k-1)/2$ independent binomial factors, one corresponding to each of the possible vectors **m** (McCullagh, 1982). The product binomial model just described was derived from a different viewpoint by Pike *et al.* (1975). The equivalent log–linear model of quasi-symmetry was investigated by Caussinus (1965) but the connection with matched pairs was not explored.

Matched triplets can be treated in a similar way (McCullagh, 1982).

5.4.4 *Discussion*

The use of some form of conditional likelihood function as a device for eliminating nuisance parameters seems necessary when the data are sparse and the number of nuisance parameters is large. Where a conditional likelihood could in principle be constructed and used for inference, the unconditional method may often nevertheless give satisfactory answers. In the example of Section 5.4.1 where the expected values in some categories were as low as 2.0 the conditional and unconditional methods differed by only 8% of a standard deviation. Of course, if information from many such tables were to be combined, the arguments in favour of the conditional method would be greatly strengthened. Matched pairs provides an extreme example where the unconditional method is quite unsatisfactory.

5.5 Bibliographic notes

Many of the methods discussed in Chapter 4 for binary data carry over to polytomous data and the references given there are also relevant here. Further details can be found in the books by Haberman (1974a), Fienberg (1980) and Plackett (1981). The development of the subject given here, which concentrates on model construction, omits a number of topics, including measures of association (Goodman and Kruskal, 1954, 1959, 1963, 1972; Agresti, 1977), tests of marginal homogeneity (Stuart, 1955; Bhapkar, 1966) and alternative strategies for fitting log–linear models. Details of the iterative proportional fitting algorithm for log–linear models can be found in Bishop *et al.* (1975). This method forms the core of some computing packages, but is not sufficiently general to cover the range of models considered here.

Tests of the hypothesis that two marginal distributions in a square table of counts are the same are given in most texts dealing with discrete data. Usually, however, such data arise from a matched-pairs design, perhaps but not necessarily from a retrospective study. The usual null hypothesis of interest, that for each pair the response probabilities are identical, would correspond to symmetry and not just marginal homogeneity. Non-null hypotheses such as those described in Section 5.4 result both in asymmetry and marginal inhomogeneity. It is readily seen that in the context of matched pairs, marginal homogeneity corresponds to a hypothesis about a null main effect in the presence of interactions involving that effect. Hypotheses that violate the marginality principle or models that are not hierarchical are rarely of scientific interest and are therefore not considered here.

The idea of representing ordered categories as contiguous intervals on a continuous scale goes back at least to Pearson (1901) who investigated coat-colour inheritance in thoroughbred horses. An extension of this idea to two variables led to the development of the tetrachoric and polychoric correlation coefficients and to the quarrel with Yule (1912) (Pearson, 1913; Pearson and Heron, 1913). The proportional-odds model described in Section 5.1.2 was used by Hewlett and Plackett (1956), Snell (1964), Simon (1974), Clayton (1974), Bock (1975) and others. Ashford (1959), Gurland *et al.* (1960) and Finney (1971) used the probit in place of the logistic function. Williams and Grizzle (1972) discuss a number of methods including the proportional-odds model and scoring methods in the log–linear context (see also Haberman, 1974a, b). McCullagh (1980) compares scoring methods with the proportional-odds model and concludes that the proportional-odds and related models based on transforming the cumulative proportions are to be preferred to scoring methods because they are invariant under the grouping of adjacent response categories.

The connection with an underlying continuous scale is lost when we deal with continuation-ratio models (Fienberg, 1980) or log–linear models with scores for the categories. Consequently such models would seem to be appropriate when categories cannot be combined without altering the essence of the measurement scale.

The need to eliminate nuisance parameters when the data are

sparse and the number of parameters large is well known and has been emphasized by Breslow (1976, 1981) in the context of binary data. The conditional method described here for the log–linear model follows the Neyman–Pearson prescription for constructing similar regions (Lehmann, 1959). The conditional method for the continuation-ratio model corresponds to partial likelihood (Cox, 1975), while the method suggested for the proportional-odds model appears to be new and is untested.

CHAPTER 6

Log–linear models

6.1 Introduction

In this chapter we are concerned mainly with counted data not in the form of proportions. Typical examples involve counts of events in a Poisson or Poisson-like process where the upper limit to the number is infinite or effectively so. One example discussed in Section 6.3 deals with the number of incidents involving damage to ships of a specified type over a given period of time. Classical examples involve radiation counts as measured in, say, particles per second by a Geiger counter. In behavioural studies counts of incidents in a time interval or specified length are often recorded.

It is clear from the form of these examples that the Poisson model for randomness in the counts will not generally be appropriate. In behavioural studies involving primates, for example, behaviour incidents generally occur in batches or clusters. Similarly, with the data on ship damage, inter-ship variability leads to over-dispersion relative to the Poisson model. Here, unless there is strong evidence to the contrary, we avoid the assumption of Poisson variation and assume only that

$$\text{var}(Y_i) = \sigma^2 \text{E}(Y_i), \qquad (6.1)$$

where σ^2, the dispersion parameter, is assumed constant over the data. Under-dispersion, a phenomenon less common in applications, is included here by putting $\sigma^2 < 1$.

In log–linear models the dependence of $\mu_i = \text{E}(Y_i)$ on the covariate vector $\mathbf{x}_i$ is assumed to be multiplicative and is usually written in the logarithmic form

$$\ln \mu_i = \eta_i = \boldsymbol{\beta}^{\text{T}}\mathbf{x}_i; \qquad i = 1, \ldots, N. \qquad (6.2)$$

When we refer to log–linear models we mean primarily the log–linear relationship (6.2); often (6.1) is tacitly assumed as a secondary aspect of the model. In applications both components of the log–linear model, but primarily (6.2), require checking.

Here, as in previous chapters, the problem of modelling reduces to finding an appropriate parsimonious model matrix $\mathbf{X}$ of order $N \times p$. As elsewhere it is important that the final model or models should make sense physically: usually this will mean that interactions should not be included without main effects nor higher-degree polynomial terms without their lower-degree relatives. Furthermore, if the model is to be used as a summary (of the findings) of one out of several studies bearing on the same phenomenon, main effects would usually be included whether significant or not.

All log–linear models have the form (6.2). Variety is created by different forms of model matrices; there is an obvious analogy with analysis-of-variance and linear regression models. In the theoretical development it is not usually necessary to specify the form of $\mathbf{X}$, though in applications, of course, the form of $\mathbf{X}$ is all-important. In Section 6.4, which deals with the connection between log–linear and multinomial response models, some aspects of the structure of $\mathbf{X}$ are important. It is shown that, under certain conditions, there is an equivalence between log–linear models and certain multinomial response models dealt with in Chapters 4 and 5.

6.2 Likelihood functions

In Chapters 4 and 5 we encountered the binomial, multinomial and hypergeometric distributions. These are appropriate as models for proportions where the total is fixed. In the present chapter we concentrate on the Poisson distribution for which the sample space is the set of non-negative integers. In particular there is no finite upper limit on the values that may be observed. The probability distribution is given by

$$\text{pr}(Y = y) = e^{-\mu}\mu^y/y!; \qquad y = 0, 1, 2, \ldots,$$

from which the cumulant generating function

$$\mu(e^t - 1)$$

may be derived. It follows that the mean, variance and all other cumulants of Y are equal to μ. It is readily shown that if all cumulants are $O(n)$ where n is some quantity tending to infinity, then

$$(Y-\mu)/\kappa_2^{\frac{1}{2}} \sim N(0, 1) + O_p(n^{-\frac{1}{2}}).$$

In particular, as $\mu \to \infty$

$$(Y-\mu)/\mu^{\frac{1}{2}} \sim N(0, 1) + O_p(\mu^{-\frac{1}{2}}).$$

This proof may also be applied to the binomial and hypergeometric distributions. For the latter, the appropriate limit is as the minimum marginal total tends to infinity.

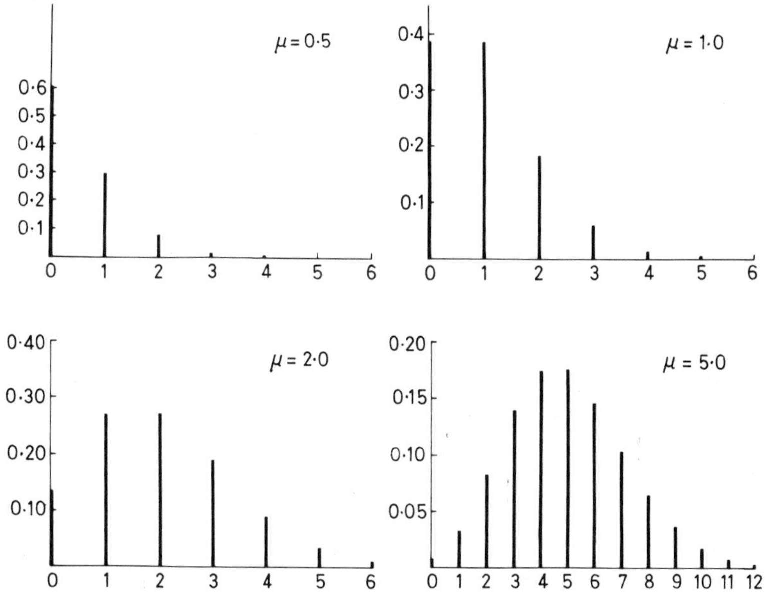

Fig. 6.1. *The Poisson distribution for* $\mu = 0.5$, 1.0, 2.0, *and* 5.0.

Sketches of the Poisson distribution are given in Fig. 6.1 for $\mu = 0.5$, 1.0, 2.0 and 5.0. These illustrate the extent of the skewness particularly for small values of μ and the approach to the Normal limit for large μ. Of course, the Normal limit refers to the cumulative distribution and not to the density.

The following properties of the Poisson distribution are sometimes useful in applications. The variance-stabilizing transform is $Y^{\frac{1}{2}}$ in the sense that for large μ

$$E(Y^{\frac{1}{2}}) \simeq \mu^{\frac{1}{2}} \qquad \text{and} \qquad \text{var}(Y^{\frac{1}{2}}) \simeq 1/4,$$

the error terms being $O(\mu^{-\frac{1}{2}})$. The power transformation to symmetry is $Y^{\frac{2}{3}}$ (Anscombe, 1953) in the sense that the standardized skewness of $Y^{\frac{2}{3}}$ is $O(\mu^{-1})$ rather than $O(\mu^{-\frac{1}{2}})$ for Y or $Y^{\frac{1}{2}}$. Neither of these transforms involves the unknown μ, although the value of μ is required when computing tail probabilities. A simple transformation producing both symmetry and stability of variance is

$$g(Y) = \begin{cases} 3Y^{\frac{1}{2}} - 3Y^{\frac{1}{6}}\mu^{\frac{1}{3}} + \mu^{-\frac{1}{2}}/6; & Y \neq 0, \\ -(2\mu)^{\frac{1}{2}} + \mu^{-\frac{1}{2}}/6; & Y = 0. \end{cases}$$

Tail probabilities may be approximated by

$$\mathrm{pr}(Y \geqslant y) \simeq 1 - \Phi(g(y - \tfrac{1}{2})),$$

with an error of order μ^{-1} rather than $O(\mu^{-\frac{1}{2}})$. For example with $\mu = 5$ we find the following:

y		7	8	9	10	11	12	13
$\mathrm{pr}(Y \geqslant y)$	(exact)	0.2378	0.1334	0.0681	0.0318	0.0137	0.0055	0.0020
	(approx.)	0.2373	0.1328	0.0678	0.0318	0.0137	0.0055	0.0021

Non-monotonicity of the function $g(y)$ is not a serious concern because, for discrete y, it occurs only if $\mu > 38$ and then in a region of negligibly small probability.

6.2.1 *The Poisson log-likelihood function*

For a single observation y the contribution to the log likelihood is $y \ln \mu - \mu$. Plots of this function versus μ, $\ln \mu$ and $\mu^{\frac{1}{3}}$ are given in Fig. 6.2 for $y = 1$. To a close approximation it can be seen that, for $y > 0$,

$$y \ln \mu - \mu = y \ln y - y - 9y^{\frac{1}{3}}(\mu^{\frac{1}{3}} - y^{\frac{1}{3}})^2/2.$$

The signed square root of twice the difference between the function and its maximum value is $3y^{\frac{1}{6}}(y^{\frac{1}{3}} - \mu^{\frac{1}{3}})$, which is the leading term in the transformation $g(y)$ above.

For a vector of independent observations the log likelihood is

$$l(\boldsymbol{\mu}, \mathbf{y}) = \Sigma\,(y_i \ln \mu_i - \mu_i), \tag{6.3}$$

and the deviance function is essentially twice the sum of squared differences

$$\begin{aligned} D(\mathbf{y}, \boldsymbol{\mu}) &= 2l(\mathbf{y}, \mathbf{y}) - 2l(\boldsymbol{\mu}, \mathbf{y}) \\ &= 2\Sigma\,[y_i \ln(y_i/\mu_i) - (y_i - \mu_i)] \\ &\simeq 9\Sigma\, y_i^{\frac{1}{3}}(y_i^{\frac{1}{3}} - \mu_i^{\frac{1}{3}})^2. \end{aligned} \tag{6.4}$$

Thus we may regard $D(\mathbf{y}; \hat{\boldsymbol{\mu}})$ as a sum of squared residuals so that maximization of the log likelihood is equivalent to minimization of

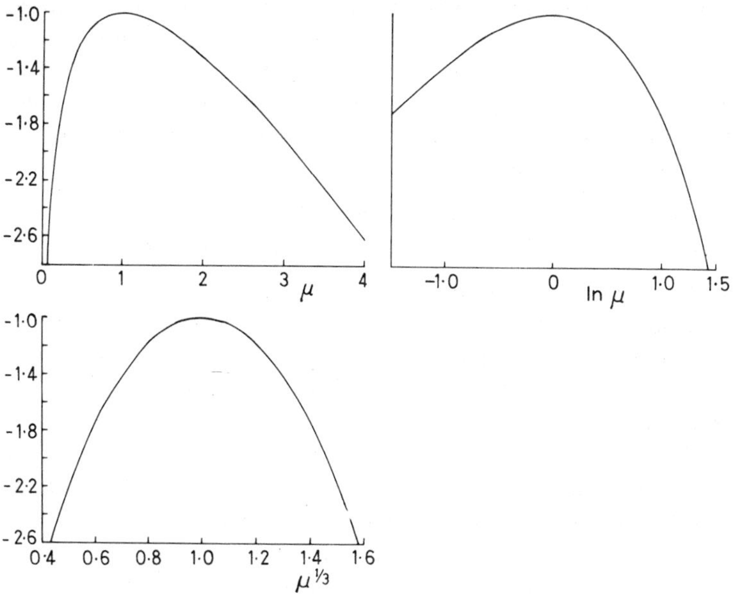

Fig. 6.2. *The Poisson log-likelihood function for* $y = 1$, *using scales* μ, *ln* μ *and* $\mu^{\frac{1}{3}}$.

$D(\mathbf{y};\boldsymbol{\mu})$. If a constant term is included in the model it is generally the case that $\Sigma(y_i - \hat{\mu}_i) = 0$ so that $D(\mathbf{y};\boldsymbol{\mu})$ may be written in the more usual form $2\Sigma\, y_i \ln(y_i/\mu_i)$.

Another approximation to $D(\mathbf{y};\boldsymbol{\mu})$ for large μ is obtained by expanding (6.4) in a Taylor series in $(y-\mu)/\mu$. We find

$$D(\mathbf{y};\boldsymbol{\mu}) \simeq \sum_i (y_i - \mu_i)^2/\mu_i,$$

which is less accurate than the quadratic approximation on the $\mu^{\frac{1}{3}}$ scale. This statistic is due to Pearson (1900).

6.2.2 *Over-dispersion*

Suppose now that the dispersion of the data is greater than that predicted by the Poisson model, i.e. $\text{var}(Y) > \text{E}(Y)$. This phenomenon can arise in a number of different ways. We might, for example, observe a Poisson process over an interval whose length is random rather than fixed. Alternatively the data might be produced by a clustered Poisson process where each event contributes a random amount to the total. In other words, we observe $Y = Z_1 + Z_2 + \ldots + Z_N$ where the Zs are independent and identically distributed and N has a Poisson distribution independent of $\mathbf{Z}$. We find for example that

$$\mathrm{E}(Y) = \mathrm{E}(N)\mathrm{E}(Z)$$

and $$\mathrm{var}(Y) = \mathrm{E}(N)\mathrm{var}(Z) + \mathrm{var}(N)[\mathrm{E}(Z)]^2 = \mathrm{E}(N)\mathrm{E}(Z^2),$$

so that there is over-dispersion if $\mathrm{E}(Z^2) > \mathrm{E}(Z)$.

Another way in which over-dispersion may arise is as follows. In behavioural studies and in studies of accident-proneness where there is inter-subject variability, the number of incidents Y for a given individual might be Poisson with mean Z. This mean is itself a random variable which might be described by the gamma distribution with mean μ and index $\phi\mu$. In other words $\mathrm{E}(Z) = \mu$, $\mathrm{var}(Z) = \mu/\phi$. This mixture leads to the negative binomial distribution

$$\mathrm{pr}(Y = y; \mu, \phi) = \frac{\Gamma(y+\phi\mu)\phi^{\phi\mu}}{y!\,\Gamma(\phi\mu)\,(1+\phi)^{y+\phi\mu}}; \qquad y = 0, 1, 2, \ldots$$

(Plackett, 1981, p. 6). The mean and variance are $\mathrm{E}(Y) = \mu$ and $\mathrm{var}(Y) = \mu(1+\phi)/\phi$. If the regression model is specified in terms of μ, say $\boldsymbol{\mu} = \boldsymbol{\mu}(\boldsymbol{\beta})$, and if ϕ is unknown but constant, then the estimating equations for $\boldsymbol{\beta}$ are in general different from those obtained by weighted least squares, and only in very simple cases do the two methods coincide. However, it may be shown that the two sets of parameter estimates, one based on the negative binomial likelihood and the other on the Poisson likelihood, differ by a term that is $O_p(\phi^{-2})$ for large ϕ. For modest amounts of over-dispersion this difference may be neglected (see also Section 10.2).

If the precise mechanism that produces the over-dispersion or under-dispersion is known (e.g. as with electronic counters), specific methods may be used. In the absence of such knowledge it is convenient to assume as an approximation that $\mathrm{var}(Y) = \sigma^2\mu$ for some constant σ^2. This assumption can and should be checked, but even relatively substantial errors in the assumed functional form of $\mathrm{var}(Y)$ generally have only a small effect on the conclusions. Parameter estimates may be obtained by maximizing the Poisson log likelihood (6.3) using, for example, the general method of Chapter 2, with the inverse matrix of second derivatives being multiplied by an estimate of σ^2 in order to obtain an appropriate measure of precision for $\hat{\boldsymbol{\beta}}$.

6.2.3 *Asymptotic theory*

The usual asymptotic results concerning, for example, consistency and asymptotic Normality of $\hat{\boldsymbol{\beta}}$ are valid provided that all elements of the information matrix increase without limit. This will generally be so if p is fixed and $N\to\infty$ or, for fixed N and p, if $\mu_i\to\infty$ for each i. The asymptotic covariance matrix of $\hat{\boldsymbol{\beta}}$ is $\sigma^2\mathbf{i}_\beta^{-1}$ where $\mathbf{i}_\beta$, the matrix of second derivatives of (6.3), emerges in a very natural way in the iterative weighted least-squares estimation procedure.

The dispersion parameter σ^2 can, if required, be estimated by

$$\tilde{\sigma}^2 = X^2/(N-p) = \sum_i \left[\frac{(y_i-\mu_i)^2}{\mu_i}\right]/(N-p).$$

By analogy with (4.14) and (5.13) the effective degrees of freedom for $\tilde{\sigma}^2$ are given by $f = (N-p)/(1+\frac{1}{2}\bar{\gamma}_2)$ so that approximate confidence limits for individual components of $\boldsymbol{\beta}$ would be based on the t_f distribution if σ^2 is unknown. This is a minor refinement and for most purposes, unless many of the means are less than 1.0, the Normal or t_{N-p} approximation is adequate.

6.3 Examples

6.3.1 *A biological assay of tuberculins*

Fisher (1949) published some data concerning a biological assay of two tuberculins, designated Standard and Weybridge, using bovine subjects. The observations are measurements in millimetres of a thickening of the skin observable in a set number of hours after intradermal injection of the tuberculin. The following is a simplified description of the experiment. One hundred and twenty cows were divided into four treatment classes, I, II, III and IV, of 30 cows each. The four treatments applied were

A Standard double,
B Standard single,
C Weybridge single,
D Weybridge half,

where 'single' refers to the amount 0.05 mg. On each cow there were four sites of application, with each cow receiving each of the four treatments. The treatment classes I, II, III and IV differed only in the sites on the neck at which the various tuberculins were applied in accordance with the following layout:

	Treatment class			
Sites	I	II	III	IV
1	A	B	C	D
2	B	A	D	C
3	C	D	A	B
4	D	C	B	A

The observations in Table 6.1 are the totals for each site and treatment class of the observed thickenings on 30 cows.

Table 6.1 *Responses in a biological assay of tuberculins*

	Treatment class				
Sites	I	III	II	IV	*Total*
1	454	249	349	249	1301
2	408	322	312	347	1389
3	523	268	411	285	1487
4	364	283	266	290	1203
Total	1749	1122	1338	1171	5380

After extensive preliminary investigation and prior to summarizing the data in the form of Table 6.1, Fisher concluded (a) that the effect of treatment and choice of site were multiplicative and (b) that the variance of any observation was roughly proportional to its expected value. Thus, although the response is a measurement, not a count, the methods of this chapter apply.

The systematic part of the model is thus log–linear where the model matrix **X** is the incidence matrix for a non-cyclic 4×4 Latin square with the treatments A, B, C and D indexed according to a 2×2 factorial arrangement with no interaction. In this way we can examine the relative potency of the two tuberculin preparations either at the high-dose level or at the low-dose level.

Parameter estimates measuring logarithmic relative potency are found by maximizing (6.3) and are similar to those obtained by Fisher (1949) who used a non-iterative method. The values are:

		Equation (6.3)	*Fisher*
B	Standard single	0.0000	0.0000
A	Standard double	0.2095	0.2089
D	Weybridge half	0.0026	0.0019
C	Weybridge single	0.2121	0.2108

Using (6.5) we find $\tilde{\sigma}^2 = 1.410/7 = 0.2014$ compared with Fisher's value of 0.2018. Taken together with the relevant components of the inverse matrix of second derivatives the standard errors for the treatment contrasts on the log scale are:

	Estimate	*SE*	*Correlation*
High vs low dose	0.2095	0.0124	−0.0053
Weybridge vs Standard	0.0026	0.0123	

Confidence limits may be constructed based on the t_7-distribution. These estimates show that the Weybridge single is slightly more potent than the Standard half dose but not significantly so, the ratio of estimated responses being exp(0.0026). Similarly, doubling the dose increases the response by an estimated factor exp(0.2095) equal to a 23.3% increase. This factor applies to both Standard and Weybridge. The relative potency of the two preparations is the ratio of weights producing equal mean responses and is estimated here as 2.017 compared with Fisher's estimate of 2.009 (which should apparently have been 2.013).

In the analysis just given it is assumed (a) that the response at one site on the neck is unaffected by the treatment applied at other sites and (b) that the effect on the logarithmic scale of doubling the dose of the Standard preparation is the same as doubling the dose of the Weybridge preparation. This later assumption can and should be checked by including in the model the interaction term between preparation and volume. This is equivalent to regarding the treatments A, B, C and D as an unstructured four-level factor instead of as two two-level factors having no interaction.

In the design of this experiment it was recognized that the variability of response between animals would be very large but that on different sites on the same animal the variability would be considerably less. It is essential, therefore, in the interests of high precision to make comparisons of the two preparations on the same animal. In the arrangement described in Table 6.1 each cow is assigned a treatment class I–IV so that contrasts between sites and treatment contrasts are within the same animal. On the other hand, contrasts between treatment classes are between animals and thus involve an additional component of dispersion. Strictly, the analysis should have been made conditional on the observed column totals but in fact, as we shall see in the following section, this would make no difference to the numerical values of the treatment contrasts or to their estimated precision. However, because of this additional component of variability, the standard errors for the treatment-class contrasts in the log–linear model that we have used are inappropriate. This complication does not invalidate the analysis given here because treatment-class contrasts are incidental.

Fisher gives a detailed discussion of the conclusions to be drawn from these data, including a study of the components of dispersion just described. The principal conclusion, that the relative potency is just in excess of 2.0, is complicated by the later discovery, using careful comparative tests with guinea-pigs, that the estimated relative potency was 0.9. Thus it would appear that the two tuberculin preparations must be qualitatively different, though such a difference is unlikely to show up in a study confined to a single species.

6.3.2 *A study of ship damage*

The data in Table 6.2, kindly provided by J. Crilley and L. N. Heminway of Lloyd's Register of Shipping, concern a type of damage caused by waves to the forward section of certain cargo-carrying vessels. For the purpose of setting standards for hull construction we need to know the risk of damage associated with the three classifying factors which are:

Ship type	A–E
Year of construction	1960–64, 1965–69, 1970–74, 1975–79
Period of operation	1960–74, 1975–79

Table 6.2 *Number of reported damage incidents and aggregate months of service by ship type, year of construction and period of operation*

Ship type	*Year of construction*	*Period of operation*	*Aggregate months service*	*Number of damage incidents*
A	1960–64	1960–74	127	0
A	1960–64	1975–79	63	0
A	1965–69	1960–74	1095	3
A	1965–69	1975–79	1095	4
A	1970–74	1960–74	1512	6
A	1970–74	1975–79	3353	18
A	1975–79	1960–74	0	0*
A	1975–79	1975–79	2244	11
B	1960–64	1960–74	44882	39
B	1960–64	1975–79	17176	29
B	1965–69	1960–74	28609	58
B	1965–69	1975–79	20370	53
B	1970–74	1960–74	7064	12
B	1970–74	1975–79	13099	44
B	1975–79	1960–74	0	0*
B	1975–79	1975–79	7117	18
C	1960–64	1960–74	1179	1
C	1960–64	1975–79	552	1
C	1965–69	1960–74	781	0
C	1965–69	1975–79	676	1
C	1970–74	1960–74	783	6
C	1970–74	1975–79	1948	2
C	1975–79	1960–74	0	0*
C	1975–79	1975–79	274	1
D	1960–64	1960–74	251	0
D	1960–64	1975–79	105	0
D	1965–69	1960–74	288	0
D	1965–69	1975–79	192	0
D	1970–74	1960–74	349	2
D	1970–74	1975–79	1208	11
D	1975–79	1960–74	0	0*
D	1975–79	1975–79	2051	4
E	1960–64	1960–74	45	0
E	1960–64	1975–79	0	0*
E	1965–69	1960–74	789	7
E	1965–69	1975–79	437	7
E	1970–74	1960–74	1157	5
E	1970–74	1975–79	2161	12
E	1975–79	1960–74	0	0*
E	1975–79	1975–79	542	1

* Necessarily empty cells.

Data courtesy of J. Crilley and L. N. Heminway, Lloyd's Register of Shipping.

The data give the number of damage incidents (as distinct from the number of ships damaged), the aggregate number of months service or total period at risk and the three classifying factors. Note that a single ship may be damaged more than once and furthermore that some ships will have been operating in both periods. No ships constructed after 1975 could have operated before 1974, explaining five of the six (necessarily) empty cells.

It seems reasonable to suppose that the number of damage incidents is directly proportional to the aggregate months service or total period of risk. This assumption can be checked later. Furthermore, multiplicative effects seem more plausible here than additive effects. These considerations lead to the initial very simple model:

$$\begin{aligned}\ln(\text{expected number of damage incidents}) &\\ = \beta_0 + \ln(\text{aggregate months service}) &\\ + (\text{effect due to ship type}) &\\ + (\text{effect due to year of construction}) &\\ + (\text{effect due to service period}). & \qquad (6.5)\end{aligned}$$

The last three terms in this model are qualitative factors. The first term following the intercept is a quantitative variate whose regression coefficient is known to be 1. Such a term is sometimes called an *offset*.

For the random variation in the model the Poisson distribution might be thought appropriate as a first approximation, but there is undoubtedly some inter-ship variability in accident-proneness. This would lead to over-dispersion as described in Section 6.2.2. Thus we assume simply that $\mathrm{var}(Y) = \sigma^2 \mathrm{E}(Y)$ and expect to find $\sigma^2 > 1$. Parameter estimates are computed using the Poisson log likelihood.

The main-effects model (6.5) fits these data reasonably well but some large residuals remain, particularly observation 21 for which the standardized residual is 2.87. Here we use the standardization $(y-\hat{\mu})/(\tilde{\sigma}\hat{\mu}^{\frac{1}{2}})$ where $\tilde{\sigma}^2 = 1.69$. As part of the standard procedure for model checking we note the following:

1. All of the main effects are highly significant.
2. The coefficient of ln(aggregate months service), when estimated, was 0.903 with approximate standard error 0.13, confirming the assumed prior value of unity.
3. Neither of the two-factor interactions involving service period is significant.
4. There is inconclusive evidence of an interaction between ship type

and year of construction, the deviance being reduced from 38.7 with 25 degrees of freedom to 14.6 with 10. This reduction would have some significance if the Poisson model were appropriate but, with over-dispersion present, the significance of the approximate F-ratio $(38.7-14.6)/(15\times 1.74)=0.92$ vanishes completely. Here, 1.74 is the estimate of σ^2 with the interaction term included.

5. Even with the interaction term included, the standardized residual for observation 21 remains high at 2.48.

We may summarize the conclusions as follows: the number of damage incidents is roughly proportional to the length of the period at risk and there is evidence of inter-ship variability ($\tilde{\sigma}^2 = 1.69$); the estimate for the effect due to service period (after vs before the 1974 oil crisis) is 0.385 with standard error 0.154. These values are virtually unaffected by the inclusion of the interaction term. Thus, on taking exponents, we see that the rate of damage incidents increased by an estimated 47% with approximate 95% confidence limits (8%, 100%) after 1974. This percentage increase applies uniformly to all ship types regardless of when they were constructed.

Ships of types B and C have the lowest risk, type E the highest. Similarly the oldest ships appear to be the safest, with those built between 1965 and 1974 having the highest risk. Parameter estimates from the main-effects model on which these conclusions are based are given in Table 6.3. Table 6.4 gives the observed rate of damage

Table 6.3 *Estimates for the main effects in the ship damage example*

Parameter		*Estimate*	*Standard error*
Intercept		−6.41	—
Ship type	A	0.00	—
	B	−0.54	0.23
	C	−0.69	0.43
	D	−0.08	0.38
	E	0.33	0.31
Year of construction	1960–64	0.00	—
	1965–69	0.70	0.19
	1970–74	0.82	0.22
	1975–79	0.45	0.30
Service period	1960–74	0.00	—
	1975–79	0.38	0.15

Table 6.4 *Rate of damage incidents* ($\times 10^3$ *per ship month at risk) by ship type and year of construction*

Ship type	*Year of construction* 1960–64	1965–69	1970–74	1975–79
A	0.0	3.2	4.9	4.9
B	1.1	2.3	2.3	2.5
C	1.2	0.7	2.9	3.6
D	0.0	0.0	8.3	2.0
E	0.0	11.4	5.1	1.8

incidents by ship type and year of construction. The reason for the suggested interaction is that the risk for ships of types A, B and C is increasing over time while the risk for type E appears to be decreasing. The above conclusions would be somewhat modified if observation 21 were set aside.

One final point concerns the computation of residual degrees of freedom for the model containing the interaction term. The method used in some computing packages gives 13 instead of 10. However, the appropriate reference set is conditional on the observed value of the sufficient statistic for the model containing the interaction term. One component of the sufficient statistic is the two-way table of marginal totals given in Table 6.4. The first three columns of this table involve sums of two observations. Apart from the four zeros which give degenerate distributions, each remaining cell in the first three columns contributes one degree of freedom, giving 11 in all. One further degree of freedom is lost because of the effect due to service period.

6.4 Log–linear models and multinomial response models

In the following sections we investigate the connection between log–linear models for frequencies and multinomial response models for proportions. The connection between these models derives from the fact that the multinomial and binomial distributions may be derived from independent Poisson random variables conditional on their observed total.

6.4.1 *Comparison of two or more Poisson means*

Suppose that $Y_1, \ldots, Y_k$ are independent Poisson random variables with means $\mu_1, \ldots, \mu_k$ and that we require to test the hypothesis $H_0: \mu_1 = \ldots = \mu_k = e^{\beta_0}$ against the alternative that, for $\beta_1, \ln \mu_j = \beta_0 + \beta_1 x_j$, where x_j are known constants. The standard theory of significance testing (Lehmann, 1959) leads to consideration of the statistic $T = \Sigma x_j Y_j$ conditional on $\Sigma Y_j = n$. In other words, in the computation of significance levels we regard the data as multinomially distributed with parameter vector $(k^{-1}, \ldots, k^{-1})$. This conditional distribution is independent of β_0 so that the significance level for alternatives $\beta_1 > 0, p^+ = \text{pr}(T \geqslant t_{\text{obs}}; H_0)$, can in principle be computed directly from the multinomial distribution. Here $t_{\text{obs}} = \Sigma x_j y_j$ is the observed value of the test statistic. Alternatively we may approximate the distribution of T by noting that under the null hypothesis T is the total of a random sample taken with replacement from the finite population $x_1 \ldots, x_k$. When $k = 2$ the above procedure gives as significance level the tail area of the binomial distribution.

The log-likelihood function for (β_0, β_1) in the above problem is

$$l_y(\beta_0, \beta_1) = \beta_0 \Sigma y_j + \beta_1 \Sigma x_j y_j - \Sigma \exp(\beta_0 + \beta_1 x_j).$$

Now make the transformation

$$\tau = \Sigma \exp(\beta_0 + \beta_1 x_j) = \exp(\beta_0) \Sigma \exp(\beta_1 x_j)$$

so that the log likelihood becomes

$$\begin{aligned} l_y(\tau, \beta_1) &= (\Sigma y_j) \ln \tau - \tau + \beta_1 \Sigma x_j y_j - \ln(\Sigma \exp(\beta_1 x_j)) \\ &= l_n(\tau; n) + l_{y|n}(\beta_1; y). \end{aligned}$$

The first component in the above expression is the Poisson log likelihood for τ as a function of n and the second is the multinomial log likelihood for β_1 as a function of $t = \Sigma x_j y_j$. All the information on β_1 resides in the second component. In particular, the information matrix for (τ, β) is diagonal so that these parameters are said to be orthogonal. Furthermore the estimate of β_1 and the estimate of its precision based on $l_{y|n}(\beta_1; y)$ are identical to those based on the Poisson log likelihood $l_y(\beta_0; \beta_1)$.

6.4.2 *Multinomial response models*

The results of the previous section may readily be generalized to demonstrate the equivalence of certain log–linear models and multinomial response models of the kind discussed in Section 5.1.3. The following discussion follows that of Palmgren (1981).

It is convenient to arrange the data y_{ij} in a two-way table, $i = 1, \ldots, N, j = 1, \ldots, k$, where the rows themselves will generally be indexed by the crossing of several factors. We suppose that the log–linear model may be written

$$\ln \mu_{ij} = \phi_i + \mathbf{x}_{ij}^{\mathrm{T}} \boldsymbol{\beta},$$

where $\mu_{ij} = \mathrm{E}(Y_{ij})$, $\boldsymbol{\beta}$ of length p is the parameter vector of interest and $\boldsymbol{\phi}$ of length N is incidental. Under this model the number of parameters is $N+p$ so that as $N \to \infty$ maximum-likelihood estimates cannot be guaranteed to have their usual asymptotically optimal properties.

The log likelihood is

$$\begin{aligned} l_y(\boldsymbol{\phi}, \boldsymbol{\beta}) &= \Sigma [y_{ij}(\phi_i + \mathbf{x}_{ij}^{\mathrm{T}} \boldsymbol{\beta}) - \exp(\phi_i + \mathbf{x}_{ij}^{\mathrm{T}} \boldsymbol{\beta})] \\ &= \sum_i \phi_i y_{i.} + \sum_{ij} y_{ij} \mathbf{x}_{ij}^{\mathrm{T}} \boldsymbol{\beta} - \sum_{ij} \exp(\phi_i + \mathbf{x}_{ij}^{\mathrm{T}} \boldsymbol{\beta}). \end{aligned}$$

Now write $n_i = y_{i.}$,

$$\tau_i = \sum_j \mu_{ij} = \sum_j \exp(\phi_i + \mathbf{x}_{ij}^{\mathrm{T}} \boldsymbol{\beta})$$

so that the log likelihood can be written

$$\begin{aligned} l_y(\boldsymbol{\tau}, \boldsymbol{\beta}) &= \sum_i (n_i \ln \tau_i - \tau_i) \\ &\quad + \sum_i \left[\sum_j y_{ij} \mathbf{x}_{ij}^{\mathrm{T}} \boldsymbol{\beta} - n_i \ln\left(\sum_i \exp(\mathbf{x}_{ij}^{\mathrm{T}} \boldsymbol{\beta}) \right) \right] \\ &= l_n(\boldsymbol{\tau}; n) + l_{y|n}(\boldsymbol{\beta}; y). \end{aligned}$$

The first term in the above expression is the Poisson log likelihood for the row totals $y_{i.}$ and the second term is a multinomial log likelihood conditional on the observed row totals $y_{i.} = n_i$. All the information concerning $\boldsymbol{\beta}$ resides in the second component. In particular, it is clear that $\hat{\boldsymbol{\beta}}$ and $\mathrm{cov}(\hat{\boldsymbol{\beta}})$ based on $l_{y|n}(\boldsymbol{\beta}; \mathbf{y})$ are identical to those based on $l_y(\boldsymbol{\phi}, \boldsymbol{\beta})$. This analysis shows that, as $N \to \infty$, $\boldsymbol{\beta}$ is consistently estimated although $\boldsymbol{\tau}$ and hence $\boldsymbol{\phi}$ are not.

The equivalence described above depends heavily on the assumption that the parameters τ_i and hence ϕ_i are unrestricted and in particular that the row totals convey no information concerning $\boldsymbol{\beta}$. This makes sense because the row totals cannot of themselves give any information concerning the ratios of the components.

To take a specific example, consider the lizard data analysed in Section 4.5. There we regarded species as the response, and treated the remaining factors H, D, S and T as explanatory. For each combination of H, D, S and T, we conditioned on the total number of lizards observed. This total cannot convey any information concerning the theoretical proportion of *Opalinus* lizards. The linear logistic model with R as response and containing the main effects of H, D, S and T is equivalent to a log–linear model with terms

$$H*D*S*T + R*(H+D+S+T).$$

It is essential here that the full four-factor interaction $H*D*S*T$ be included even though a term such as $H\cdot D\cdot S\cdot T$, say, may be formally statistically insignificant.

6.4.3 *Discussion*

When the parameter of interest is the ratio of Poisson means or, equivalently, the value of a Poisson mean as a fraction of the total, it is generally appropriate to condition on the observed total. This leads to a multinomial or binomial response model. The discussion of the previous sections shows that, as far as parameter estimates and the matrix of second derivatives are concerned, it makes no difference whether or not we condition on the totals provided that appropriate nuisance parameters are included in the log–linear model. In this respect the results here are qualitatively different from those of Sections 4.4 and 5.4, where conditioning affects both the parameter estimates and the estimate of their precision.

The principal application of this result is that it enables multinomial response models and linear logistic models to be fitted by computing packages designed primarily for log–linear models. Unfortunately the log–linear model generally contains a very large number of parameters so that methods of estimation that rely on solving systems of linear equations, where the number of equations is equal to the number of parameters, may not be very efficient. The alternative method of iterative proportional fitting (see e.g. Bishop *et al.*, 1975, p. 83) may be used instead. However, if computational facilities permit, it is generally preferable to fit the multinomial response model directly.

6.5 Multiple responses

6.5.1 *Conditional independence*

Suppose that several responses each having two or more levels are observed. We may wish to examine which of these responses are dependent on the value of other responses. This kind of analysis would generally be exploratory. Note in particular that we make no clear distinction here between responses and explanatory variables: all are considered as responses. Suppose for definiteness that three variables indexed by factors A, B and C are observed. The data thus consist of a three-way table of counts. Simple models that might explain the structure of the data are now described.

Mutual independence of the three variables is described by the log–linear model $A+B+C$. The next simplest model involving one interaction term, namely $A*B+C$, means that the joint distribution of A and B is the same at each level of C. Path models (Goodman, 1973) typically involve two or more interaction terms. For example $A*B+B*C$ implies that, conditionally on the value of B, A and C are independent. This could be interpreted as a path model, $A \to B \to C$, where A has an effect on B and B subsequently affects C but A has no direct effect on C. Of course $A*B+B*C$ is also consistent with $C \to B \to A$, so that the direction of the causal relationship is not determined by the model formula alone. To test these path models we would generally examine the $A*C$ interaction term to see whether A has an effect on C additional to that transmitted via B.

There is an important qualitative difference between log–linear models such as $A*B+B*C$ or $A*B+B*C*D+C*E$, which are interpretable in terms of conditional independence, and models such as $A*B+B*C+C*A$ or $A*B*C+B*D+C*D$, which are not. The former are said to be decomposable (Haberman, 1974a) and have closed-form maximum-likelihood estimates. All models involving one or two compound terms only are decomposable; models involving terms of degree at most 2 can be represented by graphs where the vertices are factors and the edges are two-factor terms in the model.

Decomposable models then correspond to graphs without cycles. Models involving terms of degree 3 or more can be represented by hereditary hypergraphs (Berge, 1973) but the connection between decomposability of the model and cycles in the hypergraph breaks down. Decomposable models generally have cyclic hypergraphs but the cycles are in a sense trivial because each cycle is confined to one edge. For an alternative discussion of the connection between graphs and log–linear models, see Darroch *et al.* (1980).

The above discussion suggests that it may be important to restrict attention not simply to so-called hierarchical models but to the subset of decomposable models. For example if there were p response factors, models containing more than two $(p-1)$-degree terms or more than $(p-1)$ second-degree terms would not be allowed. Often, however, only a subset of the factors may be regarded as responses and the log–linear model may be used as a technical device for fitting a multinomial response model. The definition of decomposability would now be restricted to the response variables only. For example, if A, B and C represent responses and D and E represent explanatory factors, then $A*D+D*E+E*A$ and $A*B+B*D+D*E*A$ would be regarded as decomposable but $A*B+B*C+C*A+D*E$ would not. This extension of the definition breaks the connection between decomposability and closed-form estimates for the parameters.

6.5.2 *Models for dependence*

We now consider models suitable for describing the dependence of a bivariate or multivariate response on explanatory variables $\mathbf{x}$ which may be qualitative, quantitative or a mixture of the two. For simplicity and for definiteness it is convenient to consider a bivariate binary response indexed by the response factors A and B so that each multinomial observation has four categories. The general idea is to construct three functionally independent contrasts among the four probabilities and to examine how the values of each of these constrasts depend on $\mathbf{x}$. In the case of a single multinomial observation any choice of three contrasts is equivalent to any other, but in constructing models for dependence it is important to choose a set of contrasts likely to result in a simple interpretation.

In the context of the log–linear model the contrasts that are most convenient, both mathematically and computationally, are the factorial contrasts of the logarithms of the probabilities. There is the usual problem of defining and interpreting the main-effect contrasts in the presence of an interaction. We proceed as follows. Let the probabilities and their logarithms be:

	B		total
A	Π_{11}	Π_{12}	$\Pi_{1.}$
	Π_{21}	Π_{22}	$\Pi_{2.}$
total	$\Pi_{.1}$	$\Pi_{.2}$	1

and

	B	
A	λ_{11}	λ_{12}
	λ_{21}	λ_{22}

The factorial contrasts may be written

$$\begin{array}{ll} A & \lambda_{11}-\lambda_{21}, \\ B & \lambda_{11}-\lambda_{12}, \\ A\cdot B & \lambda_{11}-\lambda_{12}-\lambda_{21}+\lambda_{22}. \end{array} \tag{6.6}$$

The term labelled A is the logistic transform of $\text{pr}(A=1\,|\,B=1)$ while $A-A\cdot B$ is the logistic transform of $\text{pr}(A=1\,|\,B=2)$. Conversely for the term labelled B.

Suppose now that there is a single quantitative explanatory variable $\mathbf{x}$. Let F be an artificially constructed factor with N levels, one for each multinomial observation. This factor is used as an artificial device to ensure that the log–linear model corresponds to a multinomial response model (Section 6.4). Various log–linear models may be fitted. In increasing order of complexity we have

$$\begin{array}{ll} \text{(I)} & A*B+F, \\ \text{(II)} & A*B+F+A\cdot X, \\ \text{(III)} & A*B+F+A\cdot X+B\cdot X, \\ \text{(IV)} & A*B+F+A\cdot X+B\cdot X+A\cdot B\cdot X. \end{array} \tag{6.7}$$

All these models imply that A and B are dependent responses. The first model states that the joint distribution is the same for all x. The second implies that the distribution of A depends additively on B and on x, while, given A, B does not depend on x. We may write

$$\begin{aligned} \text{logit}(\text{pr}(A=1\,|\,B=1)) &= \alpha_1+\beta x, \\ \text{logit}(\text{pr}(A=1\,|\,B=2)) &= \alpha_2+\beta x, \end{aligned} \tag{6.8}$$

so that the conditional probabilities for A given B lie on two parallel lines on the logistic scale. The corresponding probabilities for B given A are a pair of constants under (II) and a pair of parallel lines under

(III). The fourth model corresponds to pairs of lines that are not parallel.

Formally identical models with the same model formulae may be used when A and B have more than two levels. Again the parameters are interpreted in terms of the conditional probabilities for A given the level of B and for B given the level of A.

The above decomposition in terms of factorial contrasts of the logarithms is not necessarily always the most useful. Indeed it is difficult to think of an application involving two responses where both conditional distributions would be of interest. An alternative decomposition in terms of the marginal distributions is

$$\begin{array}{ll} A' & \log(\Pi_{1.}/\Pi_{2.}), \\ B' & \log(\Pi_{.1}/\Pi_{.2}), \\ A \cdot B & \lambda_{11}-\lambda_{12}-\lambda_{21}+\lambda_{22}. \end{array} \tag{6.9}$$

In principle at least, we could use this decomposition to construct models for dependence having formally the same structure as (6.7) but where the interaction terms with x refer to the marginal distributions of A and B rather than to the conditional distributions. Note, however, that if the conditional probabilities satisfy the linear logistic model (6.8) the marginal probabilities will have a complicated dependence on x. Conversely, though, it may well be that the marginal probabilities have a simpler structure than the conditional probabilities.

The appropriate decomposition depends to some extent on the purpose of the analysis. For example, if we wish to compare the conclusions of one study with those of another where only one variable was recorded, the decomposition (6.7) leading to a log–linear model would be inappropriate. Furthermore if any of the response variables has more than two categories it would usually be important to consider carefully the nature of the measurement scale (Section 5.1).

6.6 Bibliographic notes

For the most part the references given in Chapters 4 and 5 are also appropriate here. The books by Bishop *et al.* (1975), Goodman (1978), Haberman (1974a), Plackett (1981) and Upton (1978) are particularly relevant. Plackett gives an extensive bibliography; see also Killion and Zahn (1976). Haberman gives a mathematical account of log–linear models generally and introduces the notion of decomposability. The connection between decomposability and the theory of graphs was discussed by Darroch *et al.* (1980). Gokhale and Kullback (1978) use minimum discrimination information as the criterion for fitting log–linear models. For the most part their estimates coincide with those obtained by maximum likelihood.

CHAPTER 7

Models for data with constant coefficient of variation

7.1 Introduction

The classical linear models introduced in Chapter 3 assume that the variance of the response is a constant over the entire range of parameter values. This property is required to ensure that variance estimates are consistent and that the parameters are estimated with maximum precision by ordinary least squares. However, it is not uncommon to find data in the form of continuous measurements where the variance increases with the mean. In Chapter 6 we considered the modelling of data where $\text{var}(Y) \propto \text{E}(Y)$. Here we assume that the coefficient of variation is constant, i.e. that

$$\text{var}(Y) = \sigma^2[\text{E}(Y)]^2 = \sigma^2\mu^2,$$

where σ is the coefficient of variation. For small σ we have that

$$\text{E}(\ln Y) = \ln\mu - \sigma^2/2 \qquad \text{and} \qquad \text{var}(\ln Y) = \sigma^2.$$

Furthermore it may be that the systematic part of the model is multiplicative on the original scale and so additive on the log scale, i.e. that

$$\ln[\text{E}(Y_i)] = \eta_i = \mathbf{x}_i^{\text{T}}\boldsymbol{\beta}.$$

Then, with the exception of the intercept or constant term in the linear model, consistent estimates of the parameters and of their precision may be obtained by transforming the data and applying least squares to the vector of logarithms. The intercept is biased by an amount equal to $-\sigma^2/2$.

We may, however, choose to retain the original scale of measurement, write

$$\mu = \text{E}(Y) = \exp(\mathbf{x}^{\text{T}}\boldsymbol{\beta}),$$

use a log link function to achieve linearity, and a quadratic variance function to describe the random component. The method of iterative weighted least squares (Chapter 8) may now be used to obtain estimates for $\boldsymbol{\beta}$. This method of estimation is equivalent to assuming that Y has a gamma distribution with constant index $\nu = \sigma^{-2}$ independent of the mean. We therefore investigate the properties of the gamma distribution.

First it is helpful to compare the two methods of estimation described above. We assume that the precise distribution of Y is not specified, for if it is the comparison can be uniquely resolved by standard efficiency calculations and by considerations such as sufficiency. For example, if Y has the log-normal distribution the first method is preferred, while if it has the gamma distribution the second method is preferred. More generally, however, if Y is a variable with a physical dimension or if it is an extensive variable (Cox and Snell, 1981, p. 14) such that a sum of Ys has some well defined physical meaning, the method of analysis based on transforming to $\ln Y$ seems unappealing on scientific grounds and the second method of analysis would be preferred. Of course if the analysis is exploratory or if graphical presentation is required, transformation of the data is convenient and indeed desirable.

7.2 The gamma distribution

We write the density in the form

$$\frac{1}{\Gamma(\nu)}\left(\frac{\nu y}{\mu}\right)^{\nu} \exp\left(-\frac{\nu y}{\mu}\right) \mathrm{d}(\ln y); \qquad y \geqslant 0, \nu > 0, \mu > 0.$$

From its cumulant generating function, $-\nu \ln(1-\mu t/\nu)$, the first three moments are easily found as

$$\begin{aligned} \kappa_1 &= \mathrm{E}(Y) = \mu \\ \kappa_2 &= \mathrm{var}(Y) = \mu^2/\nu \\ \kappa_3 &= \mathrm{E}(Y-\mu)^3 = 2\mu^3/\nu^2. \end{aligned}$$

The value of ν determines the shape of the distribution. If $0 < \nu < 1$ the density has a pole at the origin and decreases monotonically as $y \to \infty$. The special case $\nu = 1$ corresponds to the exponential distribution. If $\nu > 1$ the density is zero at the origin and has a single mode at $y = \mu - \mu/\nu$; however, the density with respect to the differential element $\mathrm{d}(\ln y)$ has a maximum at $y = \mu$. Fig. 7.1 shows the form of the distribution for $\nu = 0.5$, 1.0, 2.0 and 5.0 with $\mu = 1$

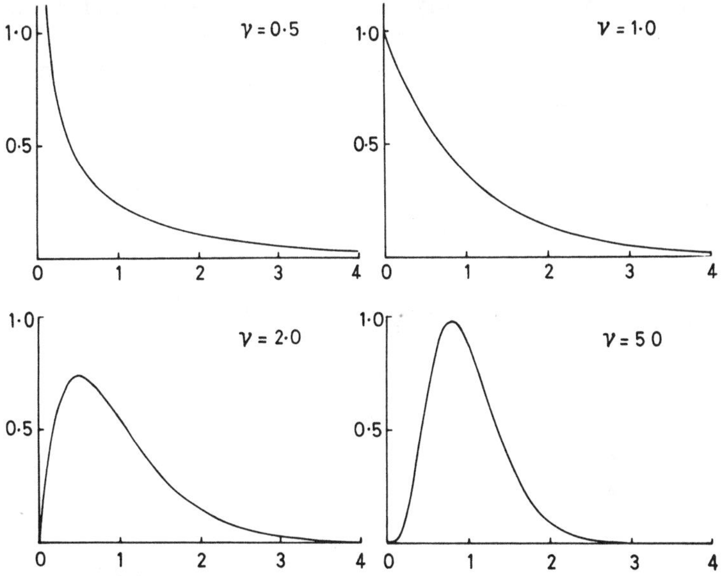

Fig. 7.1. *The gamma distribution for* $\nu = 0.5, 1.0, 2.0$ *and* 5.0, $\mu = 1$.

held constant. It can be seen from the graphs that the densities are all positively skewed. The standardized skewness coefficient is $\kappa_3/\kappa_2^{\frac{3}{2}} = 2\nu^{-\frac{1}{2}}$ and a Normal limit is attained as $\nu \to \infty$.

In this chapter we are concerned mostly with models for which the index or precision parameter $\nu = \sigma^{-2}$ is assumed constant for all observations, so that the densities all have the same shape. However, by analogy with weighted linear least squares, where the variances are proportional to known constants, we may, in the context of the gamma distribution, allow ν to vary in a similar manner from one observation to another. In other words we may have $\nu_i = \text{constant} \times w_i$, where w_i are known weights and ν_i is the index or precision parameter of Y_i. This problem occurs in components-of-variance problems where the weights are degrees of freedom of sums of squares of Normal variables and the proportionality constant is 1. Here the densities have different shapes.

The gamma family, and indeed all distributions of the type discussed in Section 2.2, are closed under convolutions. Thus if $Y_1, \ldots, Y_n$ are independent and identically distributed in the gamma distribution with index ν, then the arithmetic mean $\bar{Y}$ is distributed in the same family with index $n\nu$. Thus the gamma distribution with integer index, sometimes also called the Erlangian distribution (Cox, 1962), arises in a fairly natural way as the time to the νth event in a Poisson process.

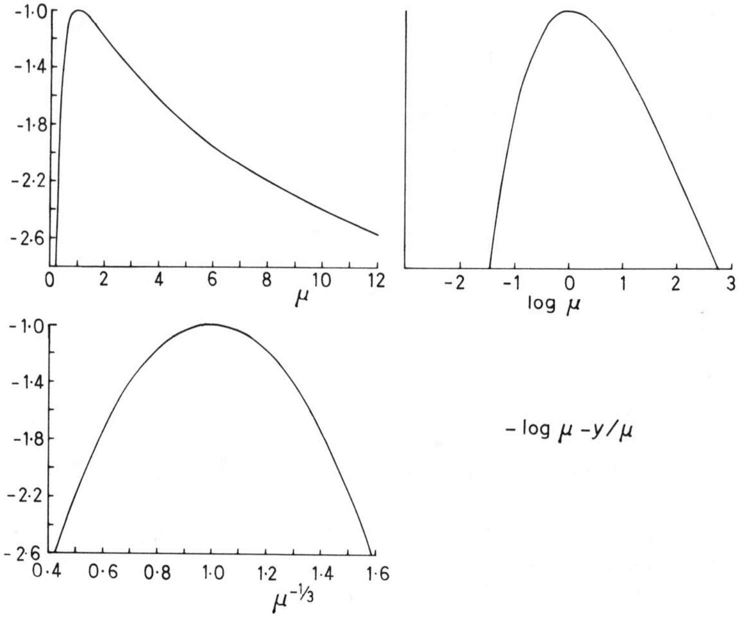

Fig. 7.2. *The gamma log likelihood plotted against μ, ln μ and $\mu^{-\frac{1}{3}}$.*

The log-likelihood function corresponding to a single observation is shown in Fig. 7.2 where we plot the log likelihood against μ, $\ln \mu$ and $\mu^{-\frac{1}{3}}$. It can be seen that the log-likelihood function is nearly quadratic on the inverse cube-root scale and that, to a close approximation, the log likelihood at μ differs from the value at the maximum by an amount

$$9y^{\frac{2}{3}}(y^{-\frac{1}{3}} - \mu^{-\frac{1}{3}})^2/2.$$

Now it is known that the square root of twice the log-likelihood ratio statistic is approximately Normally distributed. Thus a Normalizing transform for Y is

$$3[(Y/\mu)^{\frac{1}{3}} - 1].$$

The cube-root transform was previously derived in this context by Wilson and Hilferty (1931) (see also Hougaard, 1982).

7.3 Generalized linear models with gamma-distributed observations

7.3.1 *The variance function*

We have already noted that, according to the parametrization of the gamma distribution used here, the variance function is quadratic. An alternative derivation is to write the log likelihood as a function of both ν and μ in the standard form

$$\nu(-y/\mu - \ln\mu) + \nu\ln y + \nu\ln\nu - \ln\Gamma(\nu).$$

It follows that $\theta = -1/\mu$ and $b(\theta) = -\ln(-\theta)$, from which the mean $b'(\theta) = \mu$ and variance function $b''(\theta) = \mu^2$ may be derived.

7.3.2 *The deviance*

Taking ν to be a known constant, the log likelihood is

$$\sum_i \nu(-y_i/\mu_i - \ln\mu_i)$$

for independent observations. If the index is not constant but is proportional to known weights, $\nu_i = \nu w_i$, the log likelihood is equal to

$$\nu\Sigma w_i(-y_i/\mu_i - \ln\mu_i).$$

The maximum attainable log likelihood occurs at $\boldsymbol{\mu} = \mathbf{y}$ and the value attained is $-\nu\Sigma w_i(1+\ln y_i)$, which is finite unless $y_i = 0$ for some i. The deviance, which is proportional to twice the difference between the log likelihood achieved under the model and the maximum attainable value, is

$$D(\mathbf{y};\hat{\boldsymbol{\mu}}) = -2\Sigma w_i[\ln(y_i/\hat{\mu}_i) - (y_i - \hat{\mu}_i)/\hat{\mu}_i].$$

The statistic is useful only if all the observations are strictly positive. More generally, if some components of $\mathbf{y}$ are zero we may replace $D(\mathbf{y};\boldsymbol{\mu})$ by

$$D^+(\mathbf{y};\hat{\boldsymbol{\mu}}) = 2C(\mathbf{y}) + 2\Sigma w_i\ln\hat{\mu}_i + 2\Sigma w_i y_i/\hat{\mu}_i,$$

where $C(\mathbf{y})$ is an arbitrary bounded function of $\mathbf{y}$. The only advantage of $D(\mathbf{y};\hat{\boldsymbol{\mu}})$ over $D^+(\mathbf{y};\hat{\boldsymbol{\mu}})$ is that the former function is always positive and behaves like a residual sum of squares. Note, however, that the maximum-likelihood estimate of ν is a function of $D(\mathbf{y};\hat{\boldsymbol{\mu}})$ and not of $D^+(\mathbf{y};\hat{\boldsymbol{\mu}})$. Furthermore, if any component of $\mathbf{y}$ is zero then $\hat{\nu} = 0$. This is clearly not a desirable feature in most applications and alternative estimators are given in Section 7.3.6.

The final term in the expression for $D(\mathbf{y};\hat{\boldsymbol{\mu}})$ is usually identically zero and can be ignored (Nelder and Wedderburn, 1972). Under the same conditions, the final term in $D^+(\mathbf{y};\hat{\boldsymbol{\mu}})$ is equal to Σw_i and can be absorbed into $C(\mathbf{y})$.

7.3.3 *The canonical link*

The canonical link function yields sufficient statistics which are linear functions of the data and it is given by

$$\eta = \mu^{-1}.$$

Unlike the canonical links for the Poisson and binomial distributions, this transformation, which is often interpretable as the rate of a process, does not map the range of μ onto the whole real line. Thus the requirement that $\eta > 0$ implies restrictions on the βs in any linear model. Suitable precautions must be taken in computing $\hat{\boldsymbol{\beta}}$ so that negative values of $\hat{\mu}$ are avoided.

An example of the canonical link is given by the inverse polynomial response surfaces discussed by Nelder (1966). The simplest case, that of the inverse linear response, is given by

$$\eta = \beta_0 + \beta_1/x,$$

whence

$$\mu = \frac{x}{\beta_0 x + \beta_1},$$

giving a hyperbolic form for μ against x, with a slope at the origin of $1/\beta_1$ and an asymptote $1/\beta_0$. The inclusion of a term in x, so that

$$\eta = \beta_1/x + \beta_0 + \gamma_1 x,$$

is the inverse quadratic. This is shown, with the inverse linear curve, in Fig. 7.3, and again has a slope at the origin of $1/\beta_1$; μ now tends to zero for large x like $1/(\gamma_1 x)$ and the response reaches a maximum of $\beta_0 + 2\surd(\beta_1\gamma_1)$ at $x = \surd(\beta_1/\gamma_1)$. The surfaces can be generalized to include more than one covariate and by the inclusion of cross-terms in $1/x_1 x_2$, x_1/x_2, and so on. For positive values of the parameters

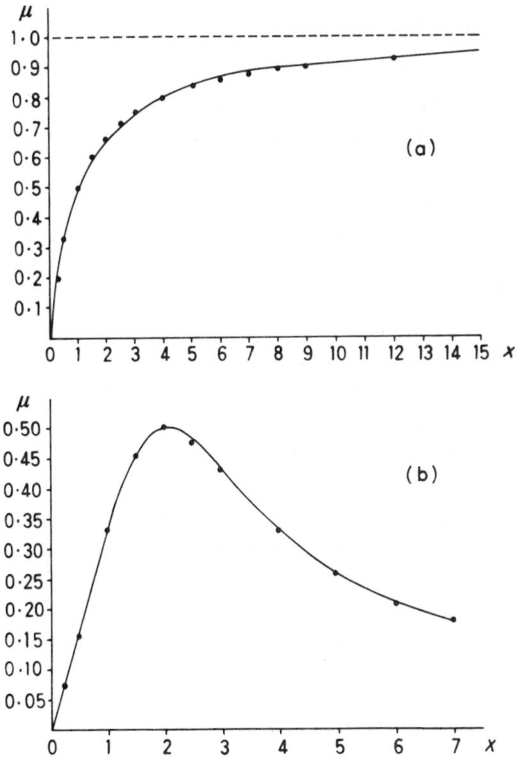

Fig. 7.3. *Inverse polynomials: (a) the inverse linear* $\mu^{-1} = 1+x^{-1}$; *(b) the inverse quadratic* $\mu^{-1} = x-2+4x^{-1}$.

the surfaces have the desirable property that the ordinate η is everywhere positive and bounded; this is in contrast to ordinary polynomials where the ordinate is unbounded at the extremes and often takes negative values.

In practice we often require to fit origins for the covariates, i.e. to make x enter the inverse polynomial in the form x_0+x, where x_0 has to be estimated. The origin x_0 is non-linear in a general sense and requires special treatment – see Chapter 10 for details.

Two other link functions are important for generalized linear models with gamma errors, the log and the identity, and we now consider their uses.

7.3.4 *Multiplicative models: log link*

We noted in Section 7.1 the close connection between linear models with constant variance for $\ln Y$ and multiplicative models with constant coefficient of variation for Y. Suppose that σ^2 is sufficiently small so that $\mathrm{var}(\ln Y) = \sigma^2 = \mathrm{var}(Y)/\mu^2$. For a linear model for $\ln Y$ the covariance matrix of the parameter estimates is $\sigma^2(\mathbf{X}^T\mathbf{X})^{-1}$, where $\mathbf{X}$ is the model matrix. For the corresponding multiplicative model the quadratic weight function is exactly unity, giving $\mathrm{cov}(\hat{\boldsymbol{\beta}}) \simeq \sigma^2(\mathbf{X}^T\mathbf{X})^{-1}$ as before. In particular, if $\mathbf{X}$ is the incidence matrix corresponding to an orthogonal design so that parameter estimates in the Normal-theory linear model are independent, then the corresponding parameter estimates in the gamma-theory multiplicative model are independent asymptotically. This property holds for all generalized linear models whenever the link function is the same as the variance-stabilizing transform.

The preceding analysis and the discussion in Section 7.1 indicate that for small σ^2 it is likely to be difficult to discriminate between Normal-theory linear models for $\ln Y$ and gamma-theory multiplicative models for Y. Atkinson's (1982) work confirms this assertion even for σ^2 as large as 0.6.

7.3.5 *Linear models: identity link*

Sums of squares of independent Normal random variables have the chi-squared or, equivalently, the gamma distribution with known index w = (degrees of freedom)/2. One method of estimating variance components is to equate the observed mean squares y_i to their expectations which are linear functions of the unknown variance components. Thus

$$\mu_i = \mathrm{E}(Y_i) = \Sigma x_{ij}\beta_j.$$

where x_{ij} are known coefficients and β_j are the variance components. Furthermore

$$\mathrm{var}(Y_i) = \mu_i^2/w_i,$$

where w_i are known weights equal to one-half the degrees of freedom of Y_i. The preceding analysis can equally well be based on sums of squares rather than on mean squares; the coefficients x_{ij} would then be replaced by $2w_i x_{ij}$ and the weights would be w_i because the coefficient of variation is unaffected by multiplication of the data by a constant.

If the number of variance components is the same as the number of mean squares, which is commonly the case, the estimating equations may be solved directly by inverting the set of linear equations. The method described above, based on the gamma likelihood, is required only when the number of independent mean squares exceeds the number of variance components. A further advantage is that approximate asymptotic variances can be obtained for the estimated variance components; unfortunately, for the higher components in particular, the size of these variances often shows that the corresponding estimates are almost worthless. Normal-theory approximations for the $\hat{\beta}$s are generally very poor.

The analysis given above is based on the assumption that the mean-square variables are independent and that the original data were Normally distributed. Furthermore, negative estimates of variance components are not explicitly ruled out; these may, however, sometimes be interpretable (Nelder, 1977).

7.3.6 *Estimation of the dispersion parameter*

The covariance matrix of the parameter estimates is $\mathrm{cov}(\hat{\boldsymbol{\beta}}) \simeq \sigma^2(\mathbf{X}^T\mathbf{W}\mathbf{X})^{-1}$, where $\mathbf{W}$ is the diagonal matrix of weights $(d\mu_i/d\eta_i)^2/V(\mu_i)$, $\mathbf{X}$ is the model matrix and σ^2 is the coefficient of variation. If σ^2 is known, this matrix may be computed directly; usually, however, it requires to be estimated. Under the gamma model the maximum-likelihood estimate of $\nu = \sigma^{-2}$ is given by

$$2N[\psi(\hat{\nu}) - \ln \hat{\nu}] = D(\mathbf{y}; \hat{\boldsymbol{\mu}}), \tag{7.1}$$

where $\psi(\nu) = \Gamma'(\nu)/\Gamma(\nu)$. A suggested improvement to take account of the fact that p parameters have been estimated is to replace the l.h.s. of the above equation by

$$2N[\psi(\hat{\nu}) - \ln \hat{\nu}] - p\hat{\nu}^{-1}, \tag{7.2}$$

the correction being the O(1) term in an asymptotic expansion for $E(D(\mathbf{Y}; \hat{\boldsymbol{\mu}}))$. There is a clear analogue here with the Normal-theory estimates of variance, $\hat{\sigma}^2$ and s^2. If ν is sufficiently large, implying σ^2 sufficiently small, we may expand (7.1) and (7.2) ignoring terms of order ν^{-2} or smaller. The maximum-likelihood estimate is then approximately

$$\hat{\nu}^{-1} = D(\mathbf{y}; \hat{\boldsymbol{\mu}})/N,$$

while the bias-corrected estimate of ν^{-1} is $D(\mathbf{y}; \hat{\boldsymbol{\mu}})/(N-p)$.

The principal problem with the maximum-likelihood estimator, and in fact with any estimator based on $D(\mathbf{y};\hat{\boldsymbol{\mu}})$, is that it is extremely sensitive to very small observations and in fact $D(\mathbf{y};\hat{\boldsymbol{\mu}})$ is infinite if any component of $\mathbf{y}$ is zero. Furthermore, if the gamma assumption is false, $\hat{\nu}^{-1}$ will not consistently estimate the coefficient of variation. For these reasons we generally prefer the moment estimator

$$\tilde{\sigma}^2 = \Sigma[(y-\hat{\mu})/\hat{\mu}]^2/(N-p) = X^2/(N-p),$$

which is consistent for σ^2, provided of course that $\boldsymbol{\beta}$ has been consistently estimated. This estimator for σ^2 may be used in the formula $\sigma^2(\mathbf{X}^T\mathbf{W}\mathbf{X})^{-1}$ to obtain an estimate of $\text{cov}(\hat{\boldsymbol{\beta}})$. It is worth mentioning that, unlike the usual Normal-theory estimator of variance s^2, the bias of $\tilde{\sigma}^2$ is $O(N^{-1})$ even if the data are distributed according to the gamma distribution. The divisor $N-p$ is preferable to N but it is not sufficient to remove the $O(N^{-1})$ bias. For a single gamma sample the expected value of $\tilde{\sigma}^2$ is

$$\sigma^2[1-\sigma^2/N+O(N^{-2})].$$

The negative bias reflects the fact that $V''(\mu) > 0$ (see Section 12.5).

7.4 Examples

7.4.1 *Car insurance claims*

The data given in Table 7.1 are taken from Baxter *et al.* (1980, Table 1) and give the average claims for damage to the owner's car for privately owned and comprehensively insured vehicles. Averages are given in pounds sterling adjusted for inflation. The number of claims on which each average is based is given in parallel. Three factors thought likely to affect the average claim are:

1. policyholder's age (PA), with eight levels, 17–20, 21–24, 25–29, 30–34, 35–39, 40–49, 50–59, 60+;
2. car group (CG), with four levels, A, B, C and D;
3. vehicle age (VA), with four levels, 0–3, 4–7, 8–9, 10+.

The numbers of claims n_{ijk} on which each average is based vary widely from zero to a maximum of 434. Since the precision of each average Y_{ijk} as measured by the variance or squared coefficient of variation is proportional to the corresponding n_{ijk}, these numbers appear as weights in the analysis. This means that the five accidentally

Table 7.1 *Average costs of claims for own damage (adjusted for inflation) for privately owned, comprehensively insured cars in 1975*

		Vehicle age							
		0–3		4–7		8–9		10+	
Policy-holders' age	*Car group*	£	No.	£	No.	£	No.	£	No.
17–20	A	289	8	282	8	133	4	160	1
	B	372	10	249	28	288	1	11	1
	C	189	9	288	13	179	1	—	0
	D	763	3	850	2	—	0	—	0
21–24	A	302	18	194	31	135	10	166	4
	B	420	59	243	96	196	13	135	3
	C	268	44	343	39	293	7	104	2
	D	407	24	320	18	205	2	—	0
25–29	A	268	56	285	55	181	17	110	12
	B	275	125	243	172	179	36	264	10
	C	334	163	274	129	208	18	150	8
	D	383	72	305	50	116	6	636	1
30–34	A	236	43	270	53	160	15	110	12
	B	259	179	226	211	161	39	107	19
	C	340	197	260	125	189	30	104	9
	D	400	104	349	55	147	8	65	2
35–39	A	207	43	129	73	157	21	113	14
	B	208	191	214	219	149	46	137	23
	C	251	210	232	131	203	32	141	8
	D	233	119	325	43	207	4	—	0
40–49	A	254	90	213	98	149	35	98	22
	B	218	380	209	434	172	97	110	59
	C	239	401	250	253	174	50	129	15
	D	387	199	299	88	325	8	137	9
50–59	A	251	69	227	120	172	42	98	35
	B	196	366	229	353	164	95	132	45
	C	268	310	250	148	175	33	152	13
	D	391	105	228	46	346	10	167	1
60+	A	264	64	198	100	167	43	144	53
	B	224	228	193	233	178	73	101	44
	C	269	183	258	103	227	20	119	6
	D	385	62	324	22	192	6	123	6

empty cells, (1, 3, 4), (1, 4, 3), (1, 4, 4), (2, 4, 4) and (5, 4, 4) for which $n = 0$, are effectively left out of the analysis.

Baxter *et al.* analyse the data using a weighted Normal-theory linear model with weights n_{ijk} and the three main effects PA+CG+VA. Here we reanalyse the data, making the assumption that the coefficient of variation rather than the variance is constant across cells. Furthermore we make the assumption that the systematic effects are linear on the reciprocal scale rather than on the untransformed scale. Justification for these choices is given in Chapters 10 and 11. The model containing main effects only may be written

$$\mu_{ijk} = \mathrm{E}(Y_{ijk}) = (\mu_0 + \alpha_i + \beta_j + \gamma_k)^{-1},$$
$$\mathrm{var}(Y_{ijk}) = \sigma^2 \mu_{ijk}^2 / n_{ijk},$$

where α_i, β_j and γ_k are the parameters corresponding to the three classifying factors PA, CG and VA. One way of interpreting the reciprocal transform is to think of μ_{ijk} as the rate at which instalments of £1 must be made to service an average claim over a fixed period of one time unit. Then μ_{ijk} is the time interval between instalments or the time purchased by an instalment of £1 in servicing an average claim in cell (i, j, k).

Table 7.2 *Goodness-of-fit statistics for a sequence of models fitted to the car insurance data* (*error, gamma; link, reciprocal*)

Model	*Deviance*	*First difference*	*d.f.*	*Mean deviance*
1	649.9			
		82.2	7	11.7
PA	567.7			
		228.3	3	76.1
PA+CG	339.4			
		214.7	3	71.6
PA+CG+VA	124.8			
		34.0	21	1.62
+PA·CG	90.7			
		19.7	21	0.94
+PA·VA	71.0			
		5.4	9	0.60
+CG·VA	65.6			
		65.6	58	1.13
Complete	0.0			

One sequence of models yielded the goodness-of-fit statistics shown in Table 7.2. Using the result that first differences of the deviance have, under the appropriate hypothesis, an approximate scaled chi-squared distribution, it is clear that the model with main effects only provides a reasonable fit and that the addition of two-factor interactions yields no further explanatory power. The estimate of $\tilde{\sigma}^2$ based on the residuals from the main-effects model is

$$\tilde{\sigma}^2 = \frac{1}{109}\Sigma n(y-\hat{\mu})^2/\hat{\mu}^2 = 1.21,$$

so that the estimated coefficient of variation is $\tilde{\sigma} = 1.1$. Estimates based on the deviance give very similar values. An examination of approximately standardized residuals using the formula $\sqrt{n}(y-\hat{\mu})/(\tilde{\sigma}\hat{\mu})$ shows the six largest residuals as corresponding to observations (2, 2, 1), (3, 2, 4), (3, 4, 4), (5, 1, 2), (5, 4, 1) and (7, 2, 1) with values 3.1, 2.9, 2.6, −2.4, −2.0 and −2.3. The positions of the cells do not show any obvious pattern, and sizes of the extreme residuals are not unusual with an effective sample size of 109.

Table 7.3 *Parameter estimates and standard errors* ($\times 10^6$) *on reciprocal scale for main effects in car insurance example*

Level	*Age group* (PA)	*Car group* (CG)	*Vehicle age* (VA)
1	0 (–)	0 (–)	0 (–)
2	101 (436)	38 (169)	366 (101)
3	350 (412)	−614 (170)	1651 (227)
4	462 (410)	−1421 (181)	4154 (442)
5	1370 (419)		
6	970 (405)		
7	916 (408)		
8	920 (416)		

Parameter estimates for the main-effects model are given in Table 7.3. Standard errors are based on the estimate $\tilde{\sigma} = 1.1$. The estimate for the intercept corresponding to all factors at their lowest level is 3410×10^{-6}. Bearing in mind that the analysis is performed here on the reciprocal scale and that a large positive parameter corresponds to a small claim, we may deduce the following. The largest average claims are made by policyholders in the youngest four age groups, i.e. up to 34, the smallest average claims by those aged 35–39 and intermediate claims by those aged 40 and over. These effects are in addition to effects due to type of vehicle and vehicle age. The value of claims decreases with car age, although not linearly. There are also marked differences between the four car groups, group D being the most expensive and group C intermediate. No significant difference is discernible between car groups A and B.

It should be pointed out that the parameter estimates given here are contrasts with level 1. In a balanced design the three sets of estimates corresponding to the three factors would be uncorrelated while the correlations within a factor would be 0.5. Even where, as here, there is considerable lack of balance the correlations do not deviate markedly from the reference values.

It is possible to test and quantify the assertions made above by fusing levels 1–4 and levels 6–8 of PA and levels 1 and 2 of CG. The deviance increases to 129.8 on 116 d.f., which is a statistically insignificant increase.

We should emphasize that the preceding analysis is not the only one possible for these data. In fact a multiplicative model corresponding to a logarithmic link function would lead to similar qualitative conclusions. As is shown in Chapter 10, the data themselves point to the reciprocal model but only marginally so, and it might be argued that quantitative conclusions would be more readily stated and understood for a multiplicative model.

7.4.2 *Clotting times of blood*

Hurn *et al.* (1945) published data on the clotting time of blood, giving clotting times in seconds (y) for normal plasma diluted to nine different percentage concentrations with prothrombin-free plasma (u); clotting was induced by two lots of thromboplastin. The data are shown in Table 7.4. A hyperbolic model for lot 1 was fitted by Bliss (1970), using an inverse transformation of the data, and for both lots 1 and 2 using untransformed data. We analyse both lots using the inverse link and gamma errors.

Initial plots suggest that a log scale for u is needed to produce inverse linearity, and that both intercepts and slopes will be different for the two lots. This is confirmed by fitting the following model sequence:

Model	*Deviance*	*d.f.*
1	7.709	17
X	1.018	16
$L+X$	0.300	15
$L+L\cdot X$	0.0294	14

Here $x = \ln u$ and L is the factor defining the lots. Clearly all the terms are necessary and the final model produces a mean deviance whose square root is 0.0458, implying approximately a 4.6% standard error on the y-scale. The two fitted lines, with standard errors for the parameters shown in parentheses, are

lot 1

$$\hat{\mu}^{-1} = -0.01655(\pm 0.00086) + 0.01534(\pm 0.00143)x$$

lot 2

$$\hat{\mu}^{-1} = -0.02391(\pm 0.00038) + 0.02360(\pm 0.00062)x$$

The plot of the Pearson residuals $(y-\hat{\mu})/\hat{\mu}$ against the linear predictor $\hat{\eta}$ is satisfactory, and certainly better than either (i) the use of constant variance for Y where the residual range decreases with $\hat{\eta}$ or (ii) the use of constant variance for $1/Y$ where the analogous plot against $\hat{\mu}$ shows the range increasing with $\hat{\mu}$. Note that constant variance for $1/Y$ implies (to the first order) $\text{var}(Y) \propto \mu^4$. Thus the assumption of gamma errors (with $\text{var}(Y) \propto \mu^2$) is 'half-way' between assuming $\text{var}(Y)$ constant and $\text{var}\,(1/Y)$ constant.

Table 7.4 *Mean clotting times in seconds (y) of blood for nine percentage concentrations of normal plasma (u) and two lots of clotting agent*

	Clotting time	
u	*Lot 1*	*Lot 2*
5	118	69
10	58	35
15	42	26
20	35	21
30	27	18
40	25	16
60	21	13
80	19	12
100	18	12

The estimates suggest that the parameters for lot 2 are a constant multiple (about 1.6) of those for lot 1. If true this would mean that $\boldsymbol{\mu}_2 = k\boldsymbol{\mu}_1$, where the suffix denotes the lot. This model, though not a generalized linear model, has simple maximum-likelihood equations for estimating α, β and k where

$$\boldsymbol{\mu}_1 = 1/\boldsymbol{\eta}_1, \qquad \boldsymbol{\eta}_1 = \alpha + \beta\mathbf{x},$$
$$\boldsymbol{\mu}_2 = k\boldsymbol{\mu}_1.$$

These are equivalent to fitting α and β to data $\mathbf{y}_1$ and $\mathbf{y}_2/k$, combined with the equation $\Sigma(y_2/\mu_1 - k) = 0$. The resulting fit gives $\hat{k} = 0.625$ with deviance $= 0.0332$ and having 15 d.f. Comparing this with the fit of separate lines gives a difference of deviance of 0.0038 on one degree of freedom against a mean deviance of 0.0021 for the more complex model. The simpler model of proportionality is not discounted, with lot 2 giving times about five-eighths those of lot 1.

7.4.3 *Modelling rainfall data using a mixture of generalized linear models*

Histograms of daily rainfall data are generally skewed to the right with a 'spike' at the origin. This form of distribution suggests that such data might be modelled in two stages, one stage being concerned with the pattern of occurrence of wet and dry days, and the other with the amount of rain falling on wet days. The first stage involves discrete data and can often be modelled by a stochastic process in which the probability of rain on day t depends on the history of the process up to day $t-1$. Often, first-order dependence corresponding to a Markov chain provides a satisfactory model. In the second stage we require a family of densities on the positive line for the quantity of rainfall. To be realistic, this family of densities should be positively skewed and should have variance increasing with μ. The gamma distribution has been found appropriate in this context.

(a) *Modelling the frequency of wet days.* Coe and Stern (1982) describe the application of generalized linear models and give references to earlier work. The data for N years form an $N \times 365$ table of rainfall amounts. (We ignore the complications introduced by leap years.) Considering the years as replicates, each of the N observations for day t is classified by the double dichotomy dry/wet and previous day dry/previous day wet. Combining over replicates we obtain, for

Table 7.5 *The 2 × 2 table of frequencies for rainfall data*

		Today		
		Wet	*Dry*	*Total*
Yesterday	*Wet*	y_0	$n_0 - y_0$	n_0
	Dry	y_1	$n_1 - y_1$	n_1
	Total	y_0	$N - y_0$	N

each of the 365 days, a 2 × 2 table of frequencies having the form of Table 7.5.

Let $\Pi_0(t)$ be the probability that day t is wet given that day $t-1$ was wet: $\Pi_1(t)$ is the corresponding probability given that day $t-1$ was dry. Ignoring end effects, the likelihood for the first-order Markov model is the product over t of terms having the form

$$\Pi_0(t)^{y_0}[1-\Pi_0(t)]^{n_0-y_0}\,\Pi_1(t)^{y_1}[1-\Pi_1(t)]^{n_1-y_1}.$$

In other words each 2 × 2 table corresponds to two independent binomial observations and the row totals are taken as fixed.

If the target parameter were the difference between $\Pi_0(t)$ and $\Pi_1(t)$, say

$$\psi(t) = \Pi_0(t)\,[1-\Pi_1(t)]/\{\Pi_1(t)\,[1-\Pi_0(t)]\},$$

it would generally be preferable to condition also on the column totals of each of the 365 tables. This would lead to a hypergeometric likelihood depending only on $\psi(t)$ (see Section 4.4.1). In this application, however, we would usually be interested in models for $\Pi_0(t)$ and $\Pi_1(t)$ themselves and not just in the difference between them. Coe and Stern use linear logistic models with various explanatory terms. Obvious choices for cyclical terms are $\sin(2\pi t/365)$, $\cos(2\pi t/365)$, $\sin(4\pi t/365)$, $\cos(4\pi t/365)$, and so on. If there is a well defined dry season, a different scale corresponding to some fraction of the year might be more appropriate. The simplest model corresponding to the first harmonic would be

$$\begin{aligned}\text{logit}(\Pi_0(t)) &= \alpha_0 + \alpha_{01}\,\sin(2\pi t/365) + \beta_{01}\cos(2\pi t/365),\\ \text{logit}(\Pi_1(t)) &= \alpha_1 + \alpha_{11}\,\sin(2\pi t/365) + \beta_{11}\cos(2\pi t/365),\end{aligned}$$

which involves six parameters. Note that if the coefficients of the harmonic terms are equal ($\alpha_{01} = \alpha_{11}$, $\beta_{01} = \beta_{11}$), the odds ratio in favour of wet days is constant over t.

Various extensions of these models are possible: a second-order model would take account of the state of the two previous days, producing four probabilities to be modelled. If it is suspected that secular trends over the years are present it is important not to regard the years as replicates. Instead, we would regard the data as $365N$ Bernoulli observations indexed by day, year and previous day wet/dry. The computational burden is increased but no new theoretical problems are involved.

(b) *Modelling the rainfall on wet days.* Coe and Stern use a multiplicative model with gamma-distributed observations to model the rainfall on wet days. The idea is to express $\ln[\mu(t)]$ as a linear function involving harmonic components. Here $\mu(t)$ is the mean rainfall on day t conditional on day t being wet. If it is assumed that no secular trends are involved, the analysis requires only the means for each day of the period with the sample sizes entering the analysis as weights. The introduction of secular trends involves the use of individual daily values in the analysis. The assumption of constant coefficient of variation requires checking. The simplest way is to group the data into intervals based on the value of $\hat{\mu}$ and to estimate the coefficient of variation in each interval. Plots against $\hat{\mu}$ should reveal any systematic departure from constancy.

(c) *Some results.* Coe and Stern present the results of fitting the models described above to data from Zinder in Niger spanning 30 years. The rainy season lasts about four months so that data are restricted to 120 days of the year. Table 7.6 shows the results of fitting separate Fourier series to $\Pi_0(t)$ and $\Pi_1(t)$ of a first-order Markov chain. Each new term adds four parameters to the model, a sine and cosine term for each Π.

Table 7.6 *Analysis of deviance for a first-order Markov chain. Rainfall data from Niger*

Model	*Deviance*	*First difference*	*d.f.*	*Mean deviance*
Intercept	483.1	–	238	
+1st harmonic	260.9	222.2	4	55.6
+2nd harmonic	235.6	25.3	4	6.3
+3rd harmonic	231.4	4.2	4	1.05
+4th harmonic	227.7	3.7	4	0.9

The mean deviance is reduced to a value close to unity, indicating a satisfactory fit, and the reductions for the third and fourth harmonic are clearly insignificant. Thus a two-harmonic model involving 10 parameters is adequate for these data.

Table 7.7 *Analysis of deviance of rainfall amounts. Data from Niger*

Model	*Deviance*	*d.f.*	*Mean deviance*
Constant	224.6	119	
+ 1st harmonic	154.5	117	
+ 2nd harmonic	147.0	115	1.28
Within days	1205.0	946	1.27

The results for the rainfall amounts are shown in Table 7.7. Again two harmonics suffice, and their inclusion reduces the between-day deviance to that within days. The size of the mean deviance indicates a distribution of the amount of rainfall on wet days is close to the exponential.

CHAPTER 8

Quasi-likelihood models

8.1 Introduction

Particular cases of quasi-likelihood functions have been encountered in previous chapters, particularly those dealing with discrete data. Essentially quasi-likelihood is equivalent to weighted least squares in the special but important case where the weights depend only on the current estimates of the regression parameters. In the following sections we describe the idea in general terms, without the requirement that the systematic part of the model should be of the generalized linear type.

8.2 Definition

Suppose that the vector of responses $\mathbf{Y}$, of length N, has mean $\boldsymbol{\mu}$ and covariance matrix $\sigma^2\mathbf{V}(\boldsymbol{\mu})$ where $\mathbf{V}(\boldsymbol{\mu})$ is a positive-semi-definite matrix whose elements are known functions of $\boldsymbol{\mu}$. In addition it is necessary to assume that the systematic part of the model is specified in terms of the mean-value parameter. Usually the systematic part will involve a model matrix $\mathbf{X}$ of order $N \times p$ but this can be absorbed into the general model

$$\boldsymbol{\mu} = \boldsymbol{\mu}(\boldsymbol{\beta}), \tag{8.1}$$

where $\boldsymbol{\beta}$ is a vector of unknown parameters required to be estimated. To ensure identifiability it is assumed that the derivative matrix $\mathbf{D} = \mathrm{d}\boldsymbol{\mu}/\mathrm{d}\boldsymbol{\beta}$ has rank p for all $\boldsymbol{\beta}$; in the case of generalized linear models this is equivalent to the assumption that $\mathbf{X}$ has full rank.

The quasi-likelihood function is considered initially as a function of $\boldsymbol{\mu}$ and is defined by the system of partial differential equations

$$\partial l(\boldsymbol{\mu};\mathbf{y})/\partial\boldsymbol{\mu} = \mathbf{V}^{-}(\boldsymbol{\mu})(\mathbf{y}-\boldsymbol{\mu}). \tag{8.2}$$

It is usually unnecessary to solve (8.2). When a solution is required it is sufficient to find functions $\boldsymbol{\theta}(\boldsymbol{\mu})$ and $b(\boldsymbol{\theta})$ satisfying $\boldsymbol{\mu} = b'(\boldsymbol{\theta})$ and to write

$$l(\boldsymbol{\mu}; \mathbf{y}) = \mathbf{y}^{\mathrm{T}}\boldsymbol{\theta} - b(\boldsymbol{\theta}) + (\text{function of } \mathbf{y}). \tag{8.3}$$

when there is no ambiguity we write $l(\boldsymbol{\beta}; \mathbf{y})$ instead of $l(\boldsymbol{\mu}(\boldsymbol{\beta}); \mathbf{y})$. The maximum quasi-likelihood equations used for determining the value of $\hat{\boldsymbol{\beta}}$,

$$\partial l(\boldsymbol{\beta}; \mathbf{y})/\partial \boldsymbol{\beta} = \mathbf{0},$$

can be written

$$\mathbf{D}^{\mathrm{T}}\mathbf{V}^{-}[\mathbf{y} - \boldsymbol{\mu}(\hat{\boldsymbol{\beta}})] = \mathbf{0}. \tag{8.4}$$

To compute $\hat{\boldsymbol{\beta}}$ it is not necessary to know either the function $l(\boldsymbol{\mu}; \mathbf{y})$ or the value of σ^2. We refer to (8.4) as least-squares equations because of the geometrical interpretation: they express the fact that the vector of residuals $\mathbf{y} - \hat{\boldsymbol{\mu}}$ is orthogonal to the columns of $\mathbf{D}$ with respect to the inner-product matrix $\mathbf{V}^{-}$, both evaluated at $\hat{\boldsymbol{\mu}}$.

8.3 Examples of quasi-likelihood functions

Apart from the multinomial case where dependence is induced by fixing the total, the most interesting class involves uncorrelated observations. Specifically we assume that

$$\mathbf{V}(\boldsymbol{\mu}) = \mathrm{diag}[V(\mu_1), \ldots, V(\mu_N)],$$

so that the quasi-likelihood involves a sum of N contributions

$$l(\boldsymbol{\mu}; \mathbf{y}) = \Sigma l(\mu_i; y_i).$$

The individual components satisfy the differential equation

$$\partial l(\mu; y)/\partial \mu = (y - \mu)/V(\mu).$$

For many simple variance functions this equation can readily be solved. Explicit solutions are given in Table 8.1. These show that many of the commonly used likelihood functions within the exponential family can be derived as quasi-likelihoods if the appropriate variance function is assumed. There is a direct analogy here with the connection between Normal-theory models and estimates obtained by minimizing a sum of squared deviations: the latter gives the same numerical values for $\hat{\beta}$ as the former, but uses only second-moment assumptions instead of full distributional assumptions.

Table 8.1. *Quasi-likelihoods associated with some simple variance functions.*

$V(\mu)$	$l(\mu; y)$	θ	*Name*	*Restrictions*
1	$(y-\mu)^2/2$	μ	Normal	–
μ	$y \log \mu - \mu$	$\log \mu$	Poisson	$\mu > 0, y \geqslant 0$
μ^2	$-y/\mu - \log \mu$	$-1/\mu$	gamma	$\mu > 0, y \geqslant 0$
μ^3	$-y/2\mu^2 + 1/\mu$	$-1/2\mu^2$	inverse Gaussian	$\mu > 0, y \geqslant 0$
μ^p	$\mu^{-p}\left(\frac{\mu y}{1-p} - \frac{\mu^2}{2-p}\right)$	$\frac{1}{(1-p)\mu^{p-1}}$	–	$\mu > 0, p \neq 0, 1, 2$
$\mu(1-\mu)$	$y \ln\left(\frac{\mu}{1-\mu}\right) + \ln(1-\mu)$	$\ln\left(\frac{\mu}{1-\mu}\right)$	binomial/n	$0 < \mu < 1, 0 \leqslant y \leqslant 1$
$\mu + \mu^2/k$	$y \ln\left(\frac{\mu}{k+\mu}\right) + k \ln\left(\frac{k}{k+\mu}\right)$	$\ln\left(\frac{\mu}{k+\mu}\right)$	negative binomial	$\mu > 0, y \geqslant 0$

Included as likelihood functions in Table 8.1 are those associated with the Normal, Poisson, binomial, negative binomial, exponential, gamma and inverse Gaussian distributions. Other exponential families include the non-central hypergeometric (Plackett, 1981, p. 30) and the multinomial distributions; these have not been included because of the complexity of their variance functions.

8.4 Properties of quasi-likelihood functions

The statistical properties of quasi-likelihood functions are similar to those of ordinary log likelihoods, except that the dispersion parameter σ^2, when it is unknown, is treated separately from $\boldsymbol{\beta}$ and is not estimated by weighted least squares. It is convenient to consider separately the properties of the score function

$$\mathbf{U}_\beta = \partial l(\boldsymbol{\beta}; \mathbf{Y})/\partial \boldsymbol{\beta} = \mathbf{D}^T\mathbf{V}^-(\mathbf{Y}-\boldsymbol{\mu}), \tag{8.5}$$

properties of $\hat{\boldsymbol{\beta}}$ and properties of the quasi-likelihood-ratio statistic. We give here a summary of the principal results; for technical details, see McCullagh (1983).

The exact mean vector and covariance matrix of $\mathbf{U}_\beta$ are $\mathbf{0}$ and $\sigma^2\mathbf{i}_\beta = \sigma^2\mathbf{D}^T\mathbf{V}^-\mathbf{D}$ respectively, where $-\mathbf{i}_\beta$ is the expected second derivative matrix of $l(\boldsymbol{\beta}; \mathbf{Y})$. Furthermore if $N^{-1}\mathbf{i}_\beta$ has a positive-definite limit as $N \to \infty$, then

$$N^{-\frac{1}{2}}\mathbf{U}_\beta \sim \mathrm{N}_p(\mathbf{0}, \sigma^2\mathbf{i}_\beta/N) + \mathrm{O}_p(N^{-\frac{1}{2}}), \tag{8.6}$$

$$\mathrm{E}(\hat{\boldsymbol{\beta}}-\boldsymbol{\beta}) = \mathrm{O}(N^{-1}) \tag{8.7}$$

and

$$N^{\frac{1}{2}}(\hat{\boldsymbol{\beta}}-\boldsymbol{\beta}) \sim \mathrm{N}_p(\mathbf{0}, N\sigma^2\mathbf{i}_\beta^{-1}) + \mathrm{O}_p(N^{-\frac{1}{2}}). \tag{8.8}$$

The notation N_p denotes the p-variate Normal distribution and the remainder terms in (8.6) and (8.8) refer to the difference between the cumulative distributions of the statistic and the Normal.

To examine the distribution of the quasi-likelihood-ratio statistic, consider two nested hypotheses H_0 and H_A where the parameter spaces are of dimension q under H_0 and $p > q$ under H_A. Denote by $\hat{\boldsymbol{\beta}}_0$ and $\hat{\boldsymbol{\beta}}$ the parameter estimates under H_0 and H_A respectively. Then under H_0

$$2l(\hat{\boldsymbol{\beta}}; \mathbf{Y}) - 2l(\hat{\boldsymbol{\beta}}_0; \mathbf{Y}) \sim \sigma^2\chi^2_{p-q} + \mathrm{O}_p(N^{-\frac{1}{2}}), \tag{8.9}$$

where χ^2_{p-q} denotes the chi-squared distribution on $p-q$ degrees of freedom. This relationship can also be expressed in terms of the deviance function with the fitted values denoted by $\hat{\boldsymbol{\mu}}_0$ and $\hat{\boldsymbol{\mu}}$ respectively. Expression (8.9) is equivalent to

$$D(\mathbf{Y};\boldsymbol{\mu}_0)-D(\mathbf{Y};\hat{\boldsymbol{\mu}}) \sim \sigma^2\chi^2_{p-q}+\mathrm{O}_p(N^{-\frac{1}{2}}).$$

As extra parameters are included in a model, the deviance function decreases in a similar way to the residual sum of squares in an ordinary linear model.

8.5 Parameter estimation

The regression parameters are obtained by solving (8.4). The Newton–Raphson method with negative second-derivative matrix replaced by its expected value $\mathbf{i}_\beta = \mathbf{D}^T\mathbf{V}^-\mathbf{D}$ results in an adjustment to $\hat{\boldsymbol{\beta}}_0$ given by

$$\hat{\boldsymbol{\beta}}_1-\hat{\boldsymbol{\beta}}_0 = (\mathbf{D}^T\mathbf{V}^-\mathbf{D})^{-1}\mathbf{D}^T\mathbf{V}^-(\mathbf{y}-\hat{\boldsymbol{\mu}}_0). \tag{8.10}$$

For generalized linear models the derivative matrix $\mathbf{D}$ is the product of a diagonal matrix $\{\mathrm{d}\mu_i/\mathrm{d}\eta_i\}$ and a model matrix $\mathbf{X}$. Furthermore $\mathbf{V}$ is a diagonal matrix with elements V_i. Denote by $\mathbf{W}$ the diagonal matrix of weights whose elements are $(\mathrm{d}\mu_i/\mathrm{d}\eta_i)^2/V_i$ and let $\mathbf{z}$ be the adjusted dependent variate whose elements are given by

$$z_i = \hat{\eta}_i+(y_i-\hat{\mu}_i)(\mathrm{d}\eta_i/\mathrm{d}\mu_i).$$

Equations (8.10) can now be written in the form more convenient for computing

$$(\mathbf{X}^T\mathbf{W}\mathbf{X})\hat{\boldsymbol{\beta}} = \mathbf{X}^T\mathbf{W}\mathbf{z}, \tag{8.11}$$

where $\mathbf{W}$ and $\mathbf{z}$ are computed at the current estimate, $\hat{\boldsymbol{\beta}}_0$.

So far we have concentrated attention on the estimation of $\boldsymbol{\beta}$. If the dispersion parameter σ^2 is unknown, an estimate is required in order to obtain numerical values for the covariance matrix of $\hat{\boldsymbol{\beta}}$ which, for generalized linear models, is approximately $\sigma^2(\mathbf{X}^T\mathbf{W}\mathbf{X})^{-1}$. The simplest estimate is

$$\tilde{\sigma}^2 = (\mathbf{y}-\hat{\boldsymbol{\mu}})^T\mathbf{V}^-(\hat{\boldsymbol{\mu}})(\mathbf{y}-\hat{\boldsymbol{\mu}})/(N-p) = X^2/(N-p), \tag{8.12}$$

where X^2 is a generalization of Pearson's statistic used in Chapters 4, 5 and 6. The error terms of (8.6), (8.8) and (8.9) are unaffected by the insertion of $\tilde{\sigma}^2$ in place of σ^2.

Other estimates of σ^2 can be based on the deviance function which is often similar to X^2. If strong distributional assumptions can be made, these estimates may be preferable to $\tilde{\sigma}^2$ but their properties are difficult to determine when the distribution of the data is not fully specified.

The variance of $\tilde{\sigma}^2$ depends on the fourth moment of the data. One approximation is

$$\mathrm{var}(\tilde{\sigma}^2) = 2\sigma^4(1+\tfrac{1}{2}\bar{\gamma}_2)/(N-p),$$

where $\bar{\gamma}_2$ is the average standardized fourth cumulant of $\mathbf{Y}$. Thus the effective degrees of freedom of $\tilde{\sigma}^2$ are given by

$$f = (N-p)/(1+\tfrac{1}{2}\bar{\gamma}_2).$$

The difficulty in applying this result lies in obtaining an approximation for $\bar{\gamma}_2$. Approximate confidence limits for individual components of β can be based on the t_f-distribution or the t_{N-p}-distribution. If f is large, these approximations differ by a negligible term.

8.6 Examples

8.6.1 *Incidence of leaf blotch on barley*

Wedderburn's (1974) example, given in Table 8.2, relates to the incidence of *Rhynchosporium secalis* or leaf blotch on 10 varieties of barley grown at nine sites in 1965. The observations give the percentage or proportion of leaf area affected; note that these are not ratios of counts of the binomial type but rather ratios of areas. An analysis using the angular transformation failed to produce either additivity of effects or stability of variance. The logistic transformation applied to the data, with some adjustment to cope with values near zero, appears to be about right both for stabilizing the variance and for producing additivity. The appropriate variance function must therefore satisfy, at least approximately, $\mathrm{var}\{\ln[Y/(1-Y)]\} = \text{constant}$ or, in other words,

$$\mathrm{var}(Y) = \sigma^2 V(\mu) = \sigma^2\mu^2(1-\mu)^2. \qquad (8.13)$$

The additive systematic part of the model may be written

$$\ln[\mu_{ij}/(1-\mu_{ij})] = \alpha_i + \beta_j, \qquad (8.14)$$

where α_i refers to the site effect and β_j to the variety effect.

The two aspects of such a model are usually arrived at after some

Table 8.2 *Incidence of* R. secalis *on leaves of 10 varieties of barley grown at nine sites; percentage leaf area affected.*

	Variety										
Site	1	2	3	4	5	6	7	8	9	10	*Mean*
1	0.05	0.00	0.00	0.10	0.25	0.05	0.50	1.30	1.50	1.50	0.52
2	0.00	0.05	0.05	0.30	0.75	0.30	3.00	7.50	1.00	12.70	2.56
3	1.25	1.25	2.50	16.60	2.50	2.50	0.00	20.00	37.50	26.25	11.03
4	2.50	0.50	0.01	3.00	2.50	0.01	25.00	55.00	5.00	40.00	13.35
5	5.50	1.00	6.00	1.10	2.50	8.00	16.50	29.50	20.00	43.50	13.36
6	1.00	5.00	5.00	5.00	5.00	5.00	10.00	5.00	50.00	75.00	16.60
7	5.00	0.10	5.00	5.00	50.00	10.00	50.00	25.00	50.00	75.00	27.51
8	5.00	10.00	5.00	5.00	25.00	75.00	50.00	75.00	75.00	75.00	40.00
9	17.50	25.00	42.50	50.00	37.50	95.00	62.50	95.00	95.00	95.00	61.58
Mean	4.20	4.77	7.34	9.57	14.00	21.76	24.17	34.81	37.22	49.33	

trial and error. Thus we might typically try several promising transforms in place of the logistic in (8.14). In the present context these might include the log–log transformation, $\ln(-\ln \mu_{ij})$, and the complementary log–log, $\ln[-\ln(1-\mu_{ij})]$, in addition to the logarithmic and angular transformations. The adequacy of the variance function (8.13) can be assessed graphically, by plotting standardized residuals $(y_{ij}-\hat{\mu}_{ij})/V^{\frac{1}{2}}(\hat{\mu}_{ij})$ against the fitted values μ_{ij} or some function of them. Details are discussed in Chapter 11.

The iterative equations (8.11) here take a particularly simple form because the weighting matrix is constant, although this fact would not usually be recognized by general-purpose computer programs. The adjusted dependent variate is given by

$$z_{ij} = \hat{\eta}_{ij} + (y_{ij}-\hat{\mu}_{ij})/[\hat{\mu}_{ij}(1-\hat{\mu}_{ij})],$$

where η_{ij} and μ_{ij} are computed on the basis of the previous fit.

The quasi-likelihood function in Wedderburn's example can be calculated explicitly in closed form. We find that $l(\boldsymbol{\mu};\mathbf{y})$ is a sum of terms each of the form

$$(2y-1)\ln\left(\frac{\mu}{1-\mu}\right)-\frac{y}{\mu}-\frac{1-y}{1-\mu}. \qquad (8.15)$$

In the notation of Chapter 3 we have

$$\theta = 2\ln\left(\frac{\mu}{1-\mu}\right)-\mu^{-1}-(1-\mu)^{-1}$$

and

$$b(\theta) = \ln\left(\frac{\mu}{1-\mu}\right)+(1-\mu)^{-1}.$$

The quasi-likelihood function (8.15) can be used in the usual way to examine the significance of varietal and site effects. However, unlike the quasi-likelihoods used in Chapter 6, the above quasi-likelihood does not appear to be an ordinary likelihood for any known distribution in the exponential family.

8.6.2 *Estimation of probabilities based on marginal frequencies*

The following example illustrates a rather simple case where estimates produced by maximizing the quasi-likelihood are different from those obtained by maximizing the likelihood. Suppose that, for each $i = 1, \ldots, N$, X_{i1} and X_{i2} are independent binomial variates

$X_{i1} \sim B(n_{i1}, \Pi_1)$, $X_{i2} \sim B(n_{i2}, \Pi_2)$. It is required to estimate the pair (Π_1, Π_2) based on the observed quantities $(n_{i1}, n_{i2}, Y_i = X_{i1}+X_{i2})$, $i = 1, \ldots, N$. Problems of this nature arise in the estimation of transition probabilities from marginal frequencies (Firth, 1982).

Under the binomial model we have that

$$E(Y_i) = n_{i1}\Pi_1 + n_{i2}\Pi_2 = \mu_i, \quad \text{say,} \tag{8.16}$$

and

$$\text{cov}(\mathbf{Y}) = \text{diag}[n_{i1}\Pi_1(1-\Pi_1)+n_{i2}\Pi_2(1-\Pi_2)].$$

It would be difficult to compute the quasi-likelihood explicitly in this case because $\text{var}(Y_i)$ is not a function of μ_i alone. However, quasi-likelihood estimates may be found in the usual way without explicitly computing the function being maximized. Because the link function in (8.16) is the identity, we may write the iterative equations in the form

$$\mathbf{X}^T \hat{\mathbf{V}}^{-1}\mathbf{X}\boldsymbol{\Pi} = \mathbf{X}^T \hat{\mathbf{V}}^{-1}\mathbf{Y},$$

where the model matrix $\mathbf{X}$ consists of the two columns corresponding to (8.16) and $\mathbf{V}$ is diagonal. Thus the probabilities are estimable only if $\mathbf{X}$ has rank 2: in other words the ratios n_{i1}/n_{i2} must not all be equal.

To take a simple artificial example, suppose we observe the following:

	$i = 1$	2	3
n_1	5	6	4
n_2	5	4	6
y	7	5	6

The quasi-likelihood estimates are $\boldsymbol{\Pi} = (0.3629,\ 0.8371)$ with approximate standard errors 0.489, 0.483 and correlation -0.948. Evidently, the sum $\Pi_1+\Pi_2$ is tolerably well estimated but there is little information concerning measures of difference, $\Pi_1-\Pi_2$, Π_1/Π_2 or $\psi = \Pi_1(1-\Pi_2)/[\Pi_2(1-\Pi_1)]$.

The likelihood function for a single observation y is

$$\sum_j \binom{n_1}{j}\binom{n_2}{y-j} \Pi_1^j(1-\Pi_1)^{n_1-j}\,\Pi_2^{y-j}(1-\Pi_2)^{n_2-y+j}.$$

The full log likelihood is a rather unpleasant expression but it can be maximized using the EM algorithm (Dempster *et al.*, 1977). We find $\hat{\Pi}_{ML} = (0.20, 1.0)$ on the boundary of the parameter space. Note that in both cases the sum of the two probabilities is estimated as 1.2, i.e. in both cases Σy is equated to $\Sigma E(Y; \hat{\Pi})$.

If all of the numbers in the above example were increased by a factor of, say, 10, the quasi-likelihood estimate of $\mathbf{\Pi}$ would be unaffected and approximate standard errors would be reduced by a factor of 0.316. However, the maximum-likelihood estimate is not similarly invariant. We find after multiplying by 10 that $\hat{\Pi}_{ML} = (0.4399, 0.7601)$. Incidentally, the EM algorithm was found to converge at an intolerably slow rate in the above exercise.

8.6.3 *Approximating the hypergeometric likelihood*

Suppose that observations arise in the form of a doubly conditioned 2×2 table distributed according to the hypergeometric distribution. Let the observations be y_{ij} with row totals n_1 and n_2, column totals m_1 and m_2 and odds ratio $\psi = e^{\beta}$. Let $Y_{11} = Y$ be the random variable corresponding to the (1, 1) element of the table. The exact first and second moments of Y are

$$\mu_{11} = E(Y) = P_1(\psi)/P_0(\psi) \qquad \text{and} \qquad E(Y^2) = P_2(\psi)/P_0(\psi),$$

where

$$P_r(\psi) = \sum_j \binom{n_1}{j}\binom{n_2}{m_1 - j}\psi^j j^r$$

is a polynomial in ψ. As usual, the sum extends from $\max(0, n_2 - m_1)$ to $\min(n_1, m_1)$. The exact log likelihood is

$$y \ln \psi - \ln P_0(\psi),$$

which is rather awkward to compute if $\min(n_1, n_2, m_1, m_2)$ is large.

Now to a close approximation we have that

$$\operatorname{var}(Y) = (n_1 + n_2)\left(\Sigma \mu_{ij}^{-1}\right)^{-1} / (n_1 + n_2 - 1) = v^*(\mu), \quad \text{say.} \tag{8.17}$$

Solving the quasi-likelihood equation and using the approximation (8.17), we find

$$l^*(\mu; y) = \frac{n_1 + n_2 - 1}{n_1 + n_2} \sum_{ij} (y_{ij} \ln \mu_{ij} - \mu_{ij}), \tag{8.18}$$

so that, apart from the factor $(n_1 + n_2 - 1)/(n_1 + n_2)$, the quasi-likelihood is the same as that for four independent Poisson or two independent binomial observations. Of course, this approximation does not avoid the problem of computing μ_{11} as a function of ψ. The exact computation involves two polynomials $P_0(\psi)$ and $P_1(\psi)$, but if this calculation should prove onerous the device described in Section 4.4.2 can be used to reduce the computations to the solution of a quadratic equation. It is important here to emphasize that μ_{11} is the conditional mean of Y given the marginal totals m_1, m_2, n_1 and n_2, as opposed to the unconditional mean treating m_1 and m_2 as random.

We now proceed to compare analytically the correct log likelihood with the approximation (8.18). A constant difference between the two would be of no consequence and so it is appropriate to compare derivatives. The correct log likelihood satisfies

$$\frac{\partial l}{\partial \mu} = \frac{y-\mu}{v(\mu)},$$

whereas the approximating log likelihood l^* satisfies

$$\frac{\partial l^*}{\partial \mu} = \frac{y-\mu}{v^*(\mu)}.$$

Furthermore $\mathrm{d}\mu/\mathrm{d}\beta = v(\mu)$ in the usual parameterization with $\beta = \ln \psi$. Thus, provided the approximation (8.17) is also used for the derivative, the two functions will have their maxima at exactly the same point. Finally, the second derivative with respect to β is a sum $\Sigma v(\mu)$ or $\Sigma v^*(\mu)$. Extensive numerical investigations show that when $v^*(\mu)$ is not exact we have $v^*(\mu) < v(\mu)$. Thus if the approximation (8.18) is used, the asymptotic variance of $\hat{\beta}$ will be slightly over-estimated.

The above discussion can readily be extended to cover the case of vector-valued parameters.

Barndorff-Nielsen and Cox (1979) give a more accurate though less convenient approximation to the hypergeometric likelihood.

To take a simple example, consider the data of Table 8.3 taken from Breslow and Day (1980, p. 145) concerning the relation between alcohol consumption and oesophageal cancer. The stratifying factor is age with six levels and there is one 2×2 table for each stratum. We consider three models, one in which the log odds ratio is zero, one in which the log odds ratio is constant and one in which the log odds ratio has a linear regression on age. The approximate deviances

$$D^*(\mathbf{y};\hat{\boldsymbol{\mu}}) = 2l^*(\mathbf{y};\mathbf{y}) - 2l^*(\hat{\boldsymbol{\mu}};\mathbf{y})$$

Table 8.3 *Data from a retrospective study concerning the relationship between alcohol consumption and oesophageal cancer*

	Cases		*Controls*			*Residual*
Age	n_1	y_{11}*	n_2	m_1†	$\hat{\mu}_{11}$	$(y_{11}-\hat{\mu}_{11})/\sqrt{v(\hat{\mu}_{11})}$
25–34	1	1	115	10	0.33	1.42
35–44	9	4	190	30	4.11	−0.07
45–54	46	25	167	54	24.49	0.18
55–64	76	42	166	69	40.09	0.60
65–74	55	19	106	37	23.75	−1.82
75+	13	5	31	5	3.24	1.75

* y_{11} = number of cases whose alcohol consumption exceeded 80 g per day.

† m_1 = total number of individuals whose alcohol consumption exceeded 80 g per day.

are 89.88, 10.77 and 10.10 on 6, 5 and 4 degrees of freedom respectively.

The estimate of $\beta = \ln \psi$ for the model of constant odds ratio is $\hat{\beta} = 1.658403$ with standard error 0.1891. If the device described in Section 4.4.2 were used for approximating μ we would find $\hat{\beta}^* = 1.658445$, showing that the approximation is perfectly adequate for statistical purposes. For the third model, which we write in the form

$$\ln \psi_i = \beta_0 + \beta_1(i - 3.5),$$

we find

$$\hat{\boldsymbol{\beta}}^* = \begin{pmatrix} 1.702668 \\ -0.125468 \end{pmatrix} \qquad \text{SE} = \begin{pmatrix} 0.2003 \\ 0.1884 \end{pmatrix}$$

using the method of Section 4.4.2 for computing μ. The exact likelihood calculations give

$$\hat{\boldsymbol{\beta}} = \begin{pmatrix} 1.702631 \\ -0.125442 \end{pmatrix} \qquad \text{SE} = \begin{pmatrix} 0.1997 \\ 0.1879 \end{pmatrix}.$$

The differences are negligible.

Thus it appears that the odds for oesophageal cancer are higher by an estimated factor of $5.251 = \exp(1.6584)$ in the alcohol group than in the non-alcohol group. Approximate 95% confidence limits for this factor are (3.625, 7.607).

An examination of the six residuals or components of deviance reveals no unexpected features. In particular there is no evidence of a trend with age; this confirms the earlier more formal analysis which found no evidence of a linear trend with age.

8.7 Bibliographic notes

The term 'quasi-likelihood' is due to Wedderburn (1974). However, the method of iterative weighted least squares, which includes quasi-likelihood, has a much longer history, particularly in dealing with over-dispersion in discrete data (Bartlett, 1936; Armitage, 1957; Finney, 1971, 1976; Williams, 1982). Healy (1981) mentions over-dispersion in a components-of-variance context involving discrete data, where the variance attributable to the contrast of interest involves the over-dispersed between-groups component. Fisher (1949), analysing the data of the example in Section 6.3.1, found $V(\mu) \propto \mu$ and used what amounted to a Poisson quasi-likelihood. Consistency, asymptotic Normality and optimality are discussed by McCullagh (1983).

CHAPTER 9

Models for survival data

9.1 Introduction

This chapter is concerned with models for the analysis of data where the dependent variate is a survival time. The data may concern parts in a machine, in which case survival time means time until breakdown, or organisms, where survival time is generally life-time, but might be time before a disease reappears. The covariates can be of the usual various types, and the data may arise from an experiment or from an observational study.

There are two characteristic aspects of survival data which deserve emphasis. The first is the widespread occurrence of *censoring*, which arises because studies are frequently ended before the outcomes of all the units are known; thus we may know only that a particular patient was still alive at a certain time, but do not know the exact survival time. This may arise because the patient withdrew from the study before it ended, or because the study itself was ended. Censoring is so common that estimation methods must allow for it if they are to be generally useful. The second characteristic of survival data is the frequent presence of *time-dependent covariates*. Such a covariate, $\mathbf{x}$, say, cannot be represented by a single value x_i for unit i, but takes values that may change with time. Thus if a patient has received some treatment in the past, represented by a factor at two levels (no = 1, yes = 2), then before the treatment was applied x has value 1 and afterwards it has value 2. The changeover may occur at different times for different patients.

9.2 Survival and hazard functions

Let the survival time for a population have a density function $f(t)$. (In practice $f(\ )$ will also depend on other parameters, but for the moment we omit reference to these.) The corresponding distribution function

$$F(t) = \int_{-\infty}^{t} f(t)\,\mathrm{d}t$$

is the fraction of the population dying by time t. The complementary function $1-F(t)$, often called the *survivor function*, is the fraction still surviving at time t. The *hazard function* $h(t)$ measures the instantaneous risk, in that $h(t)\delta t$ is the probability of dying in the next small interval δt given survival to time t. From the relation

$$\begin{aligned}&\text{pr(survival to } t+\delta t)\\ &\qquad = \text{pr(survival to } t)\,\text{pr(survival for } \delta t \,|\, \text{survival to } t)\end{aligned}$$

we have

$$1-F(t+\delta t) = [1-F(t)]\,[1-h(t)\delta t],$$

whence

$$\delta t F'(t) = [1-F(t)]h(t)\delta t,$$

so that

$$h(t) = f(t)/[1-F(t)].$$

The usefulness of a distribution for survival times will depend in part on its having a hazard function with suitable properties. Thus for large t a hazard function should not decrease, because beyond a certain point the chance of breakdown or death will generally not decrease with time. For small t various forms can be justified, including one that initially declines with t, for such a distribution could describe the behaviour of a machine part with a settling-in period, where reliability increases once the initial period is over.

The simplest hazard function, a constant, is associated with the exponential distribution of survival times and hence with the Poisson process. For if

$$f(t) = \lambda\,\mathrm{e}^{-\lambda t}; \qquad t \geqslant 0,$$

then

$$F(t) = 1-\mathrm{e}^{-\lambda t},$$

and so

$$h(t) = \lambda.$$

Other forms of hazard function will appear later.

9.3 Proportional-hazards models

The hazard function depends in general both on time and on a set of covariates, the values of which may on occasion themselves depend on time. The proportional-hazards model separates these components by specifying that the hazard at time t for an individual whose covariate vector is $\mathbf{x}$ is given by

$$h(t;\mathbf{x}) = \lambda(t)\exp[G(\mathbf{x};\boldsymbol{\beta})],$$

where the second term is written in exponential form because it is required to be positive. This model implies that the ratio of the hazards for two individuals is constant over time provided that the covariates do not change. It is conventional, but not necessary (Oakes, 1981), to assume that the effects of the covariates on the hazard are also multiplicative. This additional assumption leads to models which may be written

$$h(t;\mathbf{x}) = \lambda(t)\exp(\boldsymbol{\beta}^{\mathrm{T}}\mathbf{x}), \tag{9.1}$$

where $\boldsymbol{\eta} = \boldsymbol{\beta}^{\mathrm{T}}\mathbf{x}$ is the linear predictor. The model thus implies that the ratio of hazards for two individuals depends on the difference of their linear predictors at any time, and so, with no time-dependent covariates, is a constant independent of time. This is a strong assumption and will clearly need checking in applications. Various assumptions may be made about the $\lambda(t)$ function. If a continuous survival distribution is assumed, $\lambda(t)$ will be a smooth function of t, defined for all t. Cox's model (Cox, 1972) treats $\lambda(t)$ as analogous to the block factor in a blocked experiment, defined only at points where deaths occur, thus making no assumptions about the trend with time. In practice it frequently makes surprisingly little difference to estimates and inferences whether we put a structure on the λ-function or not. We consider first estimation of $\boldsymbol{\beta}$ in the linear predictor with an explicit survival distribution, following closely the development in Aitkin and Clayton (1980).

9.4 Estimation with censored survival data and a survival distribution

We begin with the proportional-hazards model in the form (9.1). We require to develop the likelihood for the data, some of which may be censored, and for this we need both the density function and the survivor function.

Now

$$h(t) = F'(t)/[1-F(t)] = \lambda(t)\,e^{\eta},$$

so that

$$-\ln[1-F(t)] = \Lambda(t)\,e^{\eta},$$

where

$$\Lambda(t) = \int_{-\infty}^{t} \lambda(u)\,du.$$

Thus the survivor function is given by

$$S(t) = 1-F(t) = \exp[-\Lambda(t)\,e^{\eta}],$$

and the density function by minus its derivative, i.e.

$$f(t) = \lambda(t)\exp[\eta-\Lambda(t)\,e^{\eta}].$$

At the end of the study an individual who died at time t contributes a factor $f(t)$ to the likelihood, while one censored at time t contributes $S(t)$. Suppose now that we define w as a variate taking the value 1 for an uncensored observation and value 0 for a censored one, and let there be n uncensored and m censored observations. Then the log likelihood takes the form

$$\begin{aligned} l &= \sum_{i=1}^{n+m} [w_i \ln f(t_i) + (1-w_i)\ln S(t_i)] \\ &= \sum_i \{w_i[\ln \lambda(t_i)+\eta_i] - \Lambda(t_i)\,e^{\eta_i}\} \\ &= \sum_i \left[w_i[\ln \Lambda(t_i)+\eta_i] - \Lambda(t_i)\,e^{\eta_i} + w_i \ln\left(\frac{\lambda(t_i)}{\Lambda(t_i)}\right)\right]. \end{aligned}$$

Now if we write $\mu_i = \Lambda(t_i)\,e^{\eta_i}$, l becomes

$$\sum_i (w_i \ln \mu_i - \mu_i) + \sum_i w_i \ln\left(\frac{\lambda(t_i)}{\Lambda(t_i)}\right).$$

The first term is identical to the kernel of the likelihood function for $(n+m)$ independent Poisson variates w_i with means μ_i, while the second term does not depend on the unknown βs. Thus, given $\Lambda(t)$, we can obtain the estimates of the βs by treating the censoring indicator variate w_i as Poisson distributed with mean $\mu_i = \Lambda(t_i)\,e^{\eta_i}$. The link function is the same as for log–linear models except that there is a fixed intercept $\ln \Lambda(t_i)$ to be included in the linear predictor. Such a quantity is known in GLIM terminology as an *offset*.

The estimation process will not in general be quite straightforward,

the offset may contain parameters of the survival
ıes will not be known in advance. Before considering
; we deal first with the exponential distribution for
ies arise.

ential distribution

For this distribution $\lambda(t)$ becomes the constant λ so that

$$\Lambda(t) = \int_0^t \lambda(t)\,\mathrm{d}t = \lambda t.$$

Thus $\lambda(t)/\Lambda(t) = 1/t$ and no extra parameters are involved. It follows that

$$\ln \mu_i = \ln t_i + \eta_i,$$

so that the offset is just $\ln t_i$ and the log–linear model can be fitted directly. Two other distributions give particularly simple forms for $\Lambda(t)$ and these we now consider.

9.4.2 *The Weibull distribution*

By setting $\Lambda(t) = t^\alpha$, $\alpha > 0$, we obtain a hazard function proportional to $\alpha t^{\alpha-1}$ and a corresponding density $f(t)$ in the Weibull form:

$$f(t) = \alpha t^{\alpha-1}[\exp(\eta - t^\alpha e^\eta)]; \qquad t \geqslant 0.$$

Now $\lambda(t)/\Lambda(t) = \alpha/t$ and depends on the unknown parameter α, which must be jointly estimated with the βs. The kernel of the log-likelihood function is

$$n \ln \alpha + \sum_i (w_i \ln \mu_i - \mu_i),$$

and, given α, the likelihood equations for the βs are the same as those for a log–linear model with offset $\alpha \log t_i$. The equation for α given the βs takes the form

$$n/\hat{\alpha} = \sum_i (\hat{\mu}_i - w_i)\ln t_i. \tag{9.2}$$

The estimation procedure begins with $\alpha = 1$ (the exponential distribution), uses the log–linear model algorithm to fit the βs, then estimates α from (9.2), oscillating between the two stages until convergence is attained. Aitkin and Clayton (1980) note that, if α_1

and α_2 are two successive estimates of α, then convergence is faster if one uses $(\alpha_1+\alpha_2)/2$, rather than α_2, as the next working value.

Note that the log likelihood for a model differs from that of the log–linear model by the inclusion of the extra term $n\ln\hat{\alpha}$, so that the deviance requires adjustment by a term $-2n\ln\hat{\alpha}$.

9.4.3 *The extreme-value distribution*

This distribution corresponds to $\Lambda(t) = e^{\alpha t}$, giving a hazard function proportional to $\alpha e^{\alpha t}$ and a density $f(t)$ in the extreme-value form

$$f(t) = \alpha e^{\alpha t} \exp(\eta - e^{\alpha t+\eta}). \tag{9.3}$$

Note that the transformation $u = e^t$ transforms the distribution to the Weibull form. It follows that we need only replace t by u in the estimating procedure for the Weibull to obtain the corresponding one for this distribution.

9.4.4 *Other distributions*

The methods described above can, in principle, be applied to any survival distribution. Aitkin and Clayton (1980) describe a generalized extreme-value distribution which includes the Weibull and the standard extreme-value distribution as special cases. For other survival distributions the functions $\lambda(t)$ and $\Lambda(t)$ may be more awkward to compute, but there is no essential difference in fitting models which use them.

9.5 Example

9.5.1 *Remission of leukaemia*

The data in Table 9.1 from Freireich *et al.* (1963) have been analysed by Gehan (1965), Aitkin and Clayton (1980) and others. There are two samples of 21 patients each, sample 1 having been given an experimental drug and sample 2 a placebo. The times of remission are in weeks and figures in parentheses denote censored observations. To fit a survival model of the type discussed in Section 9.4, we set up a pseudo-Poisson variable taking values 0 for the censored and 1 for the uncensored observations. To fit the exponential distribution

Table 9.1 *Times of remission (weeks) of leukaemia patients, treated with drug (sample 1) and placebo (sample 2)*

Sample 1	(6)	6	6	6	7	(9)	(10)
	10	(11)	13	16	(17)	(19)	(20)
	22	23	(25)	(32)	(32)	(34)	(35)
Sample 2	1	1	2	2	3	4	4
	5	5	8	8	8	8	11
	11	12	12	15	17	22	23

Data from Freireich *et al.* (1963).
Figures in parentheses denote censored observations.

we apply an offset of $\log t$, and models with a single mean and with separate sample means (S) give deviances as follows:

Model	*Deviance*	*d.f.*
1	54.50	41
S	38.02	40

The fit is satisfactory, but it is of interest to try the more general Weibull distribution, which yields for model S the value $\hat{\alpha} = 1.366$ with a deviance of 34.13 on 39 d.f. The reduction in deviance of 3.89 is thus marginally significant at the 5% level. Separate fits of α to the two samples give similar estimates of 1.35 and 1.37; the estimate of the sample difference for $\hat{\alpha} = 1.366$ is $b_1 = 1.731$, corresponding to a hazard ratio of $\exp(1.731) = 5.65$ for sample 2 as compared with sample 1. The standard error of b_1, for α with a fixed prior value of 1.366, is ± 0.398; to adjust this for the simultaneous fitting of α, we must border the information matrix for the two parameters in the linear predictor with the second derivatives that include α and then invert the expanded matrix. The details for this example are given in Aitkin and Clayton (1980) and give $b_1 = 1.73 \pm 0.41$; the validity of this SE may be checked by plotting the deviance for fixed values of β_1 in the region of the estimate 1.73. The resulting curve rises slightly more steeply on the lower than on the upper side, but the effect at the $2 \times \text{SE}$ distance is quite small. Thus our 95% limits for the log hazard difference are $1.73 \pm (1.96 \times 0.41) = (0.93, 2.53)$ corresponding to hazard ratios of (2.52, 12.6).

Table 9.2 *Comparison of estimators for the leukaemia data*

Model (treatment of ties)	b_1	*SE*
Exponential	1.53	0.40
Weibull	1.73	0.41
Cox (Peto)	1.51	0.41
Cox (Cox)	1.63	0.43

The data of this example have been analysed by Whitehead (1980) using Cox's model and treating ties by both Peto's and Cox's methods. His results (after correcting observation 6 in sample 1, which was censored), together with those obtained above using parametric survival functions, are summarized in Table 9.2. The estimates all fall within a range of about half a standard error, and the increase in standard error from the Cox model as against the parametric survival functions is quite small. Efron (1977) and Oakes (1977) discuss this phenomenon from a theoretical viewpoint. It should be noted that the data are in units of a week, and so quite heavily grouped at the shorter times; this grouping has not been allowed for in the fits using the parametric survival functions. However, replacement of the values $t = 1$ in the first two cases of sample 2, where the grouping is most severe, by combinations of 0.5 and 1.5 has little effect on the estimates of α and β_1 in the Weibull model; thus the grouping effect is unlikely to be serious.

9.6 Cox's proportional-hazards model

Cox (1972) introduced a version of the proportional-hazards model that is only partially parametric in the sense that the baseline hazard function $\lambda(t)$ is not modelled as a smooth function of t. Instead, $\lambda(t)$ is permitted to take arbitrary values and is irrelevant in the sense that it does not enter into the estimating equations derived from Cox's partial likelihood (Cox, 1975).

First observe that we need only consider times at which failures occur because, in principle at least, the hazard could be zero over intervals that are free of failures and no contribution to the likelihood would be made by these intervals. Let $t_1 < t_2 < \ldots$ be the distinct failure times and suppose for simplicity that there are no ties in failure times. The risk set immediately prior to the jth failure, $R(t_j)$, is the set of individuals any of whom may be found to fail at time

t_j. Individuals who have previously failed or who have been censored are therefore excluded from $R(t_j)$. Given that one failure is to occur in the interval $(t_j - \delta t, t_j)$ the probability of failure for each of the individuals in $R(t_j)$ is proportional to the value of their corresponding hazard function. Let $\mathbf{x}_j$ be the value of the covariate vector for the failed individual. The probability under the proportional-hazards model that the individual who fails at time t_j is the one actually observed is

$$\frac{\lambda(t)\exp(\boldsymbol{\beta}^{\mathrm{T}}\mathbf{x}_j)}{\Sigma\,\lambda(t)\exp(\boldsymbol{\beta}^{\mathrm{T}}\mathbf{x})} = \frac{\exp(\boldsymbol{\beta}^{\mathrm{T}}\mathbf{x}_j)}{\Sigma\exp(\boldsymbol{\beta}^{\mathrm{T}}\mathbf{x})}, \tag{9.4}$$

where summation is over the risk set $R(t_j)$.

This conditional probability is the probability of observing $\mathbf{x}_j$ in sampling from the finite population corresponding to the covariate vectors in $R(t_j)$, where the selection probabilities are proportional to $\exp(\boldsymbol{\beta}^{\mathrm{T}}\mathbf{x})$. This is a generalization of the non-central hypergeometric distribution (4.21). Effectively the argument given above reverses the roles of random failure times and fixed covariates to fixed failure times and covariates selected according to the probability distribution described above.

The partial likelihood for $\boldsymbol{\beta}$, which does not involve the baseline hazard function $\lambda(t)$, is the product over the failure times of the conditional probabilities (9.4). These conditional probabilities have the form of a linear exponential-family model so that the estimate of β can be found by equating the vector sum of the covariates of the failed individuals to the sum of their conditional means. Note however that the conditioning event changes from one failure time to the next as individuals are removed from the risk set either through failure or through censoring.

9.6.1 *The treatment of ties*

The occurrence of ties among the failure times complicates the analysis, and several techniques have been proposed for dealing with this complication. One method due to Cox is as follows. Suppose for definiteness that two failures occur at time t and that the vector sum of the covariates of these two failed individuals is $\mathbf{s}_j$. The factor corresponding to (9.4) is then defined to be

$$\exp(\boldsymbol{\beta}^{\mathrm{T}}\mathbf{s}_j)/\Sigma\exp(\boldsymbol{\beta}^{\mathrm{T}}\mathbf{s}), \tag{9.5}$$

where the sum in the denominator is over all distinct pairs of individuals in $R(t_j)$. In other words we construct the finite population consisting of sums of the covariate vectors for all distinct pairs of individuals in the risk set at time t_j. The probability under an exponentially weighted sampling scheme that the failures were those of the pair actually observed is given by (9.5) which also has the exponential-family form. Note however that the number of terms in the denominator of (9.5) quickly becomes exceedingly large for even a moderate number of ties at any failure time.

Any reasonable method for dealing with ties, such as that suggested above, is likely to be satisfactory if the number of failed individuals constitutes only a small fraction of the risk set. In fact the likelihood contribution (9.5) is exact only if failures are thought of as occurring in discrete time. In applications, however, ties occur principally because of grouping. With grouped data the appropriate likelihood (Peto, 1972) involves the sum over all permutations of the failed individuals consistent with the ties observed. Suppose, for example, that two failures are tied and that the failed individuals have covariate vectors $\mathbf{x}_1$ and $\mathbf{x}_2$. The probability for the sequence in time $(\mathbf{x}_1, \mathbf{x}_2)$ or $(\mathbf{x}_2, \mathbf{x}_1)$, either of which is possible given the tie, is

$$\frac{\exp(\boldsymbol{\beta}^{\mathrm{T}}\mathbf{x}_1)}{\Sigma_R \exp(\boldsymbol{\beta}^{\mathrm{T}}\mathbf{x})} \frac{\exp(\boldsymbol{\beta}^{\mathrm{T}}\mathbf{x}_2)}{\Sigma_{R_1} \exp(\boldsymbol{\beta}^{\mathrm{T}}\mathbf{x})} + \frac{\exp(\boldsymbol{\beta}^{\mathrm{T}}\mathbf{x}_2)}{\Sigma_R \exp(\boldsymbol{\beta}^{\mathrm{T}}\mathbf{x})} \frac{\exp(\boldsymbol{\beta}^{\mathrm{T}}\mathbf{x}_1)}{\Sigma_{R_2} \exp(\boldsymbol{\beta}^{\mathrm{T}}\mathbf{x})}, \qquad (9.6)$$

where R_j is the risk set excluding $\mathbf{x}_j$ ($j = 1, 2$). Clearly the likelihood contribution becomes increasingly cumbersome as the number of ties becomes appreciable.

Expressions (9.5) and (9.6) for the contribution to the likelihood can both be thought of in terms of exponentially weighted sampling from a finite population without replacement. If the number of ties is small we may use the simpler expression

$$\frac{\exp(\boldsymbol{\beta}^{\mathrm{T}}\mathbf{s})}{[\Sigma_R \exp(\boldsymbol{\beta}^{\mathrm{T}}\mathbf{x})]^m}, \qquad (9.7)$$

where $\mathbf{s}$ is the sum of the covariate vectors of the m tied individuals (Peto, 1972). This factor corresponds to sampling with replacement.

9.6.2 *Numerical methods*

The likelihood formed by taking the product over failure times of the conditional probabilities (9.4) can, in principle, be maximized directly using the weighted least-squares method discussed in Chapters

2 and 8. Alternatively we can regard the covariate vector of the failed individuals as the response and condition on the set of covariates of all individuals in the risk set at each failure time, these being regarded as fixed. If we write $\mathbf{y}$ for the covariate vector of the failed individual the log likelihood for one failure time takes the form

$$\boldsymbol{\beta}^{\mathrm{T}}\mathbf{y} - \ln\left[\Sigma \exp(\boldsymbol{\beta}^{\mathrm{T}}\mathbf{x})\right],$$

with summation over the risk set. This has the form of an exponential family with canonical parameter $\boldsymbol{\beta}$ and $b(\theta)$ (in the notation of Section 2.2) equal to $\ln[\Sigma \exp(\boldsymbol{\beta}^{\mathrm{T}}\mathbf{x})]$. The (conditional) mean is then given by $b'(\theta)$ and the variance by $b''(\theta)$. However, this formulation is unhelpful computationally because there is no explicit expression for the quadratic weight (here equal to the variance function) as a function of the mean.

The computational difficulty can be avoided by a device similar to that used in Section 9.4. Suppose that k_j individuals are at risk immediately prior to t_j and that just one individual is about to fail. If we regard the observation on the failed individual as a multinomial observation with k_j categories, taking the value 1 for the failed observation and 0 for the remainder, then the contribution to the likelihood is again of the form (9.4), but now interpreted as a log–linear model for the cell probabilities. Thus the numerical methods of Chapter 5 may be used provided that the algorithm allows variable numbers of categories for the multinomial observations.

Alternatively (Whitehead, 1980) a Poisson log likelihood may be used provided that a blocking factor associated with failure times is included. The idea here is that at each failure time each individual in the risk set contributes an artificial Poisson response of 1 for failure and 0 for survival. The mean of this response is $\exp(\alpha + \boldsymbol{\beta}^{\mathrm{T}}\mathbf{x})$ for an individual whose covariate value is $\mathbf{x}$ and α represents the blocking factor associated with failure times. Because of the equivalence of the Poisson and multinomial likelihoods discussed in Section 6.4, the estimate of $\boldsymbol{\beta}$ and the estimate of its precision are identical to those obtained from the multinomial likelihood and hence to the partial likelihood.

The computations can be simplified if the number of distinct covariate vectors is small so that individuals in the risk set may be grouped into sets of constant hazard. The adjustment for ties is simple for the third method described above (often called Peto's method). In the multinomial log likelihood we set the multinomial total equal to the observed number of tied failures at that time. No adjustment to the algorithm is required. The corresponding Poisson log likelihood is equivalent to Peto's version of the partial likelihood.

Whitehead (1980) describes the adjustments to the Poisson likelihood required to maximize the likelihood corresponding to Cox's method for dealing with ties.

9.7 Bibliographic notes

The new large literature on the analysis of survival data includes books by Elandt-Johnson and Johnson (1980), Gross and Clark (1975), Lawless (1982), Lee (1980), Kalbfleisch and Prentice (1980) and Miller (1981). Cox's model was proposed in Cox (1972), and fitting via GLIM discussed by Whitehead (1980); the pseudo-Poisson model for parametric survival functions was proposed by Aitkin and Clayton (1980), who also discuss the definition of residuals and the necessary adaptation of standard graphical techniques (see also Crowley and Hu, 1977). For a comparison of Cox and Weibull models, see Byar (1983).

CHAPTER 10

Models with additional non-linear parameters

10.1 Introduction

Some nomenclature requires to be established first. The word 'non-linear' in the chapter title is being used in a generalized sense because, apart from the subset of classical linear models (Chapter 3), all generalized linear models are in a strict sense non-linear. However, their non-linearity is constrained in that it enters only through the variance function and the link function, the linearity of components contributing to the linear predictor being preserved. So far we have assumed that the variance function is a known function of the mean, except possibly for a dispersion parameter, and that the link function is also a known function. In this chapter we describe some generalized linear models in which unknown parameters enter either the variance or the link functions, and we also consider the use of terms in the linear predictor of a non-linear type. We shall refer to such parameters as 'non-linear' in this general sense.

The introduction of non-linear parameters complicates the fitting algorithm either by introducing an extra level of iteration or by introducing covariates which change with each iteration. The introduction of the latter may render convergence of the iterative process much less certain, and may also require starting values to be given for the non-linear parameters. In addition asymptotic covariances for the linear terms may be produced by the algorithm conditional on fixed values of the non-linear parameters, and so need adjustment when the non-linear parameters are themselves to be estimated.

10.2 Parameters in the variance function

In the generalized linear models described in Chapters 3–7, four distributions for error were involved: two of them, the Normal and gamma, contained dispersion parameters which required to be estimated, for the calculation of (asymptotic) covariances for the parameter estimates; while the other two, the Poisson and binomial, did not, although here quasi-likelihood arguments extended the analysis to include an unknown dispersion parameter also. Knowledge of the dispersion parameter is not required for solution of the likelihood equations, and in that sense it occupies a special place.

The negative binomial distribution provides an example of an unknown parameter in the variance function which is not a dispersion parameter. The distribution, which is discrete, takes the form

$$pr(Y = y;\ \alpha,\ k) = \frac{(y+k-1)!}{y!(k-1)!}\,\frac{\alpha^y}{(1+\alpha)^{y+k}}; \qquad y = 0, 1, 2\ldots.$$

The mean and variance are given by

$$\mathrm{E}(Y) = \mu = k\alpha,$$

$$\mathrm{var}(Y) = k\alpha + k\alpha^2 = \mu + \mu^2/k.$$

The log likelihood can be written in the form

$$l = y\ln[\alpha/(1+\alpha)] - k\ln(1+\alpha) + (\text{function of } y, k),$$

which has the form of a generalized linear model for fixed k, with canonical link

$$\eta = \ln\left(\frac{\alpha}{1+\alpha}\right) = \ln\left(\frac{\mu}{\mu+k}\right)$$

and variance function

$$V = \mu + \mu^2/k,$$

where the term μ can be thought of as the Poisson variance function and μ^2/k as the extra component arising from mixing the Poisson distribution with a gamma distribution for the mean to obtain the negative binomial. In general k will not be known *a priori*, and is clearly not a dispersion parameter. Estimates of k for single samples and for several samples have been discussed by Anscombe (1949), but we require an estimator for arbitrarily structured data, and a possible estimator is that which makes the mean deviance equal to unity. As with the gamma distribution an alternative estimator may be obtained by choosing k to make the Pearson X^2 statistic

$$\Sigma\frac{(y-\hat{\mu})^2}{V(\hat{\mu})}$$

equal to its expectation. The maximum-likelihood estimator is not convenient to use, requiring as it does the log gamma function and its derivative. Note that because k enters explicitly into the maximum-likelihood equations the asymptotic covariances of the parameters as given by the standard algorithm should strictly be adjusted to allow for the estimation of k.

Little use seems to have been made of the negative binomial distribution in applications; in particular the use of the canonical link is problematical. See, however, Manton *et al.* (1981) for an analysis of survey data in which both α and k are made to depend upon the classifying covariates; such a model, though of undoubted interest, lies outside the present framework.

For another example of additional parameters in the variance function, using quasi-likelihood, consider data which are to be modelled with gamma errors, but which have been collected with an absolute measurement (rounding) error, rather than with the desirable proportional error. The latter would allow the total error structure to retain the form $V = \sigma^2\mu^2$, but with the former we have a variance function of the form $V = \tau^2 + \sigma^2\mu^2$, where the first term describes the component of error arising from the constant rounding error and the second the assumed underlying gamma errors. The effect of this modified variance function is to reduce relatively the weight given to small observations. The quasi-likelihood model with this variance function would require the estimation of σ^2/τ^2 in the same way that k must be estimated for models using the negative binomial distribution.

10.3 Parameters in the link function

While link functions are often assumed known in fitting generalized linear models, it may be useful to assume on occasion that the link function comes from a class of functions, members of the class being indexed by one or more unknown parameters. The goodness of fit expressed as a function of these parameters can then be inspected to see what range of parameter values is consistent with the data. If a particular value is of interest we can perform a goodness-of-link test (Pregibon, 1980) by comparing the deviance for that value with the deviance for the best-fitting value or by using a score test.

10.3.1 *One link parameter*

A commonly considered class of link functions is that given by the power function, either in the form

$$\begin{aligned} \eta &= \mu^\theta; \qquad && \theta \neq 0, \\ \eta &= \ln \mu; && \theta = 0, \end{aligned}$$

or in the form having continuity at $\theta = 0$,

$$\eta = \frac{\mu^\theta - 1}{\theta}.$$

Exploration of this class of functions used as transformations of the data rather than of the fitted values was considered by Box and Cox (1964). The relation of the deviance to choice of θ can be explored by giving θ a set of values and evaluating the deviance accordingly. If we wish to optimize over θ we can adopt the linearizing strategy proposed by Pregibon (1980), whereby we expand the link function in a Taylor series about a fixed θ_0 and take only the linear term. Thus for the power family we have

$$\begin{aligned} g(\mu;\, \theta) = \mu^\theta &\simeq g(\mu;\, \theta_0) + (\theta - \theta_0) g'_\theta(\mu; \theta_0) \\ &= \mu^{\theta_0} + (\theta - \theta_0)\mu^{\theta_0} \ln \mu, \end{aligned} \tag{10.1}$$

so that we can approximate the correct link function

$$\eta = \eta^\theta$$

by

$$\begin{aligned} \eta_0 = \mu^{\theta_0} &= \mu^\theta - (\theta - \theta_0)\mu^{\theta_0} \ln \mu \\ &= \Sigma \beta_i x_i - (\theta - \theta_0)\mu^{\theta_0} \ln \mu. \end{aligned}$$

Now given a first estimate θ_0 of θ we can fit the model by including an extra covariate $-\mu^{\theta_0} \ln \mu$ in the model matrix, whose parameter estimate measures $\theta - \theta_0$, the first-order adjustment to θ_0. The reduction in deviance given by the inclusion of this extra covariate (using for its values fitted values $\hat{\mu}$ obtained from fitting the model without it) gives a test of θ_0 as an acceptable value for θ. To obtain the maximum-likelihood estimate for θ we have to repeat the above process forming a new adjusted value for θ at each stage. Convergence is not guaranteed, however, and requires that θ_0, our starting value, be sufficiently close to θ for the linear expansion (10.1) to be adequate. To obtain convergence, an inner iteration, whereby the extra covariate's values are refined for fixed θ_0, may be needed. Pregibon (1980) comments 'that the method is likely to be most useful in determining if a reasonable fit can be improved, rather than the somewhat more optimistic goal of correcting a hopeless situation'.

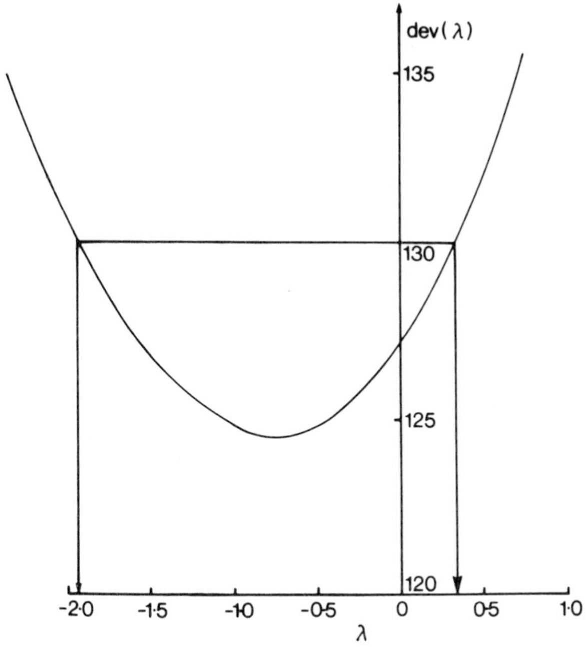

Fig. 10.1. *Car insurance data: deviance for varying λ in the link function $\eta = \mu^\lambda$ with 95% confidence limits. Main effects in linear predictor and variance function $\propto \mu^2$.*

Fig. 10.1 shows the effect with the car insurance data of changing the link by varying λ in the power family $\eta = \mu^\lambda$. The linear predictor contains the main effects only and the variance function is given by $V(\mu) \propto \mu^2$. The minimum deviance is close to $\lambda = -1$, corresponding to the reciprocal link originally chosen for the analysis, though the 95% limits show an appreciable range of compatible values for λ, including zero, corresponding to the log link.

10.3.2 *More than one link parameter*

The above method extends in principle to more than one parameter in the link function. For each parameter θ, we add an extra covariate

$$-\left(\frac{\partial g}{\partial \theta}\right)_{\theta=\theta_0}$$

to the model matrix and its parameter estimate gives the first-order adjustment to the starting value θ_0. Pregibon (1980) discusses two examples with two parameters; the first is given by

$$g(\mu;\alpha,\delta) = [(\mu+\alpha)^{\delta}-1]/\delta,$$

i.e. a form of the power family, indexed by δ, but with an added unknown origin α. Note that

$$g(\mu;1,1) = \mu,$$

so that the identity link is a member of the family.

The second of Pregibon's examples is useful for models based on tolerance distributions, such as probit analysis. The generalized link function is given by

$$g(\mu;\alpha,\delta) = \frac{\Pi^{\alpha-\delta}-1}{\alpha-\delta} - \frac{(1-\Pi)^{\alpha+\delta}-1}{\alpha+\delta}$$

when Π is the proportion responding, i.e. μ/n. The family contains the logit link as the limiting form

$$\lim_{\alpha,\delta\to 0} g(\mu;\alpha,\delta).$$

The one-parameter link family for binomial data,

$$g(\mu;\alpha) = \ln\{[(1/(1-\Pi))^{\alpha}-1]/\alpha\},$$

contains both the logistic ($\alpha = 1$) and complementary log–log link ($\alpha \to 0$) as special cases. This family may be used to assess the adequacy of an assumed linear logistic model.

10.3.3 *Transformations of data vs transformations of fitted values*

Transformations of fitted values through link functions (whether with or without unknown parameters) must be distinguished from transformations of the data values. The latter are fully discussed in Box and Cox (1964), who deal in particular with the power family.

In using a function $g(y)$, rather than y, in the analysis we are usually seeking a transformation that will yield *simultaneously* additive effects in the systematic part of the model and constancy of variance for the random part. Such a search may be successful; see, for example, the data set on survival times, given by Box and Cox, where a reciprocal transformation of the data allowed a classical regression model to be applied. However, there is no guarantee that both properties will result from the same transformation. Thus Nelder and Wedderburn (1972) in their reanalysis of the tuberculin-test data of Fisher (1949) (see the example in Section 6.3.1) show that while a square-root transformation produces desirable error properties, a log transformation is required for additivity of effects. It is an advantage of generalized linear models over other methods that the transformation to produce additivity can be made through the link function quite independently of any transformation of the data to produce approximate Normality. Indeed the latter is itself often rendered unnecessary by the possibility of using a variance function other than a constant. Thus with the Fisher data mentioned above analysis of

$$Y \text{ with variance function } V \propto \mu$$

and of

$$Y^{\frac{1}{2}} \text{ with variance function } V = \text{constant}$$

produce effectively identical results (Baker and Nelder, 1978, Appendix D).

10.4 Non-linear parameters in the covariates

As remarked in Section 3.3.1 a function of x, such as e^{kx}, is an acceptable covariate in a linear predictor, provided that k is known; we simply use the values of e^{kx} in place of x in the model matrix. However, if k requires to be estimated from the data, then non-linearity arises. Box and Tidwell (1962) describe a fitting technique by linearization which follows closely that for non-linear parameters in the link function described above. Again if $g(x;\theta)$ is the covariate to be used, with θ unknown, we expand about an initial value θ_0 to give the linear approximation

$$g(x;\theta) \approx g(x;\theta_0) + (\theta-\theta_0)\,[\partial g/\partial\theta]_{\theta=\theta_0}.$$

Thus if a non-linear term in the linear predictor is given by

$$\beta g(x;\theta),$$

we replace it by two linear terms

$$\beta u + \gamma v,$$

where

$$u = g(x; \theta_0),$$
$$v = [\partial g / \partial \theta]_{\theta = \theta_0},$$
$$\gamma = \beta(\theta - \theta_0).$$

Again an extra level of iteration is required, and after fitting a generalized linear model with u and v as covariates we obtain

$$\theta_1 = \theta_0 + \hat{\gamma}/\hat{\beta}$$

as the improved estimate, and iterate to convergence. Again, convergence is not guaranteed for starting values arbitrarily far from the solution. If the process does converge then the presence of the extra term γv will ensure that the asymptotic covariances produced for the remaining parameters are correctly adjusted for the fitting of θ. If we wish to obtain the asymptotic variance of $\hat{\theta}$ directly, we need a final iteration with $\hat{\beta} v$ in place of v; then the elements corresponding to the covariate in the $(\mathbf{X}^T\mathbf{W}\mathbf{X})^-$ matrix will give the variance of $\hat{\theta}$ and its covariance with the other parameters.

While this technique is undoubtedly useful, and indeed probably under-used, it will often be unwise to try to include more than a very few non-linear parameters in this way, especially when the other covariates are themselves appreciably correlated in the data set. It will usually be found that estimates of the non-linear parameters will have large sampling errors, and be highly correlated with the linear parameters and perhaps with each other. This is especially likely to be so in models where the systematic part consists of sums of exponentials of the form

$$\beta_0 + \beta_1 e^{k_1 x_1} + \beta_2 e^{k_2 x_2},$$

with the ks in the exponents requiring to be estimated as well as the βs.

An example where non-linear parameters arise in a fairly natural way concerns models for the joint action of a mixture of drugs (see the example in Section 10.5.3). Here, apart from a single parameter that enters the model non-linearly, the model is of the generalized linear type with covariate $\ln(x_1 + \theta x_2)$, where x_1 and x_2 are the amounts of the two drugs in the mixture. One method of analysis is to use the linearizing technique described above. Alternatively, following Darby and Ellis (1976), we may maximize the likelihood for various values of θ and plot the residual sum of squares RSS(μ) against μ. The minimum, usually unique, gives $\hat{\theta}$ and a residual mean deviance $s^2 = D(\hat{\mu})/(n-p)$ where p is the total number of parameters, including θ. Approximate confidence limits for θ can be found from

$$\{\theta: \mathrm{RSS}(\theta) - \mathrm{RSS}(\hat{\theta}) < s^2 F^*_{1,\,n-p,\,\alpha}\},$$

where $F^*_{1,\,n-p,\,\alpha}$ is the upper $100(1-\alpha)$ percentage point of the F-distribution on 1 and $n-p$ degrees of freedom. Unfortunately this method of analysis does not allow the covariances of the parameter estimates to be calculated easily. If these are required, the linearization technique should be used (see example in Section 10.5.3).

10.5 Examples

10.5.1 *The effects of fertilizers on coastal Bermuda grass*

Welch *et al.* (1963) published the results of a 4^3 factorial experiment with the three major plant nutrients, nitrogen (N), phosphorus (P) and potassium (K), on the yield of coastal Bermuda grass. The four levels for the three factors (all in lb/acre) were:

N	0	100	200	400
P	0	22	44	88
K	0	42	84	168

The yields (in tons/acre) averaged over three years and three replicates are shown in Table 10.1 with the factor levels coded 0, 1, 2, 3. Inspection of the data shows a hyperbolic type of response to the nutrients but the yield for the (0, 0, 0) plot shows that it will be necessary to allow for the nutrients already present in the soil if inverse polynomials (Section 7.3.3) are to be used to describe the response surface. We consider, therefore, an inverse linear response surface with

$$1/\mu = \eta = \beta_0 + \beta_1 u_1 + \beta_2 u_2 + \beta_3 u_3,$$

where $u_i = 1/(x_i + \alpha_i)$, $i = 1, 2, 3$. Here x_i ($i = 1, 2, 3$) are the applied amounts of N, P and K respectively, while α_i are the (unknown) amounts in the soil. If we assume for the moment that the proportional

Table 10.1 *Yields of coastal Bermuda grass as affected by N, P and K*

Nitrogen (*N*)	*Phosphorus* (*P*)	*Potassium* (*K*) 0	1	2	3
0	0	1.98	2.13	2.19	1.97
0	1	2.38	2.24	2.10	2.60
0	2	2.18	2.56	2.22	2.47
0	3	2.22	2.47	2.94	2.48
1	0	3.88	3.91	3.66	4.07
1	1	4.35	4.59	4.47	4.55
1	2	4.14	4.36	4.55	4.35
1	3	4.26	4.72	4.83	4.85
2	0	4.40	4.91	5.10	5.23
2	1	5.01	5.64	5.68	5.60
2	2	4.77	5.69	5.80	6.07
2	3	5.17	5.45	5.85	6.43
3	0	4.43	5.31	5.15	5.87
3	1	4.95	6.27	6.49	6.54
3	2	5.22	6.27	6.35	6.72
3	3	5.66	6.24	7.11	7.32

Data from Welch *et al.* (1963).

standard deviation of the yield is constant, we have model with gamma errors and the reciprocal (canonical) link, but with three non-linear parameters α_1, α_2 and α_3 to be estimated. The linearizing technique of Section 10.4 leads to our adding the extra covariates $v_i = \partial u_i/\partial \alpha_i = -u_i^2$ to the model, giving the corrections

$$\delta a_i = c_i/b_i; \qquad i = 1, 2, 3,$$

to the current estimates of α_i in each iteration, where a_i is the estimate of α_i, c_i is the coefficient of v_i and b_i that of u_i. Starting values are required for a_i and these can be obtained by taking reciprocals of the data, forming the N, P and K margins and plotting these against u_i for various trial values of α_i. The following suggested values are obtained

$$a_1 = 40, \qquad a_2 = 22, \qquad a_3 = 32.$$

Six iterations refine these to

$$a_1 = 44.60, \qquad a_2 = 15.56, \qquad a_3 = 32.39,$$

with a final deviance of 0.1965 with 57d.f., corresponding to a percentage SE per observation of 5.9. The X^2 statistic is 0.1986, trivially different from the deviance as is typical when the coefficient of variation is small. A final iteration with v_i replaced by $b_i v_i$ enables us to obtain the asymptotic standard errors of the a_i directly. The parameter estimates with their standard errors, which agree closely with those obtained from an approximate non-iterative method by Nelder (1966), are given by:

b_0	0.09746 ± 0.00963
b_1	13.5 ± 1.350
b_2	0.7007 ± 0.457
b_3	1.336 ± 0.956
a_1	44.6 ± 4.18
a_2	15.6 ± 8.44
a_3	32.4 ± 19.1

A feature of the correlation matrix is the high correlations between the a_i and b_i. These are

$$0.9702, \qquad 0.9850, \qquad 0.9849,$$

and reflect the fact that if the a_i are taken as known the standard errors of the b_i are reduced by factors of from 4 to 6. Note too the large standard errors for a_2 and a_3, which do not exclude impossible negative values; the assumptions of (asymptotic) Normality must be treated cautiously here.

The inverse linearity of the response may be tested by including the inverse quadratic terms $(x_i + \alpha_i)$ in the model. This gives a deviance of 0.1938 with 54 d.f., a negligible reduction. The Pearson residuals $(y - \hat{\mu})/\hat{\mu}$ show one possible outlier, the yield 2.94 for levels (0, 3, 2). The fitted value is 2.43, so that the possibility of a transposition at some stage from 2.49 to 2.94 might be investigated. Omission of this point does not change the fit greatly, the largest effect being on b_2. A plot of the residuals against fitted values does not contradict the assumption of gamma errors.

As shown in Nelder (1966), a quadratic polynomial with 10 parameters fits less well than an inverse linear surface with unknown origins having seven parameters. The latter is also additive for the three nutrients whereas the quadratic polynomial requires all the two-factor interactions terms for an adequate fit.

Table 10.2 *Data from assay on insecticide and synergist*

Number killed, r	*Sample size,* n	*Dose of insecticide*	*Dose of synergist*
7	100	4	0
59	200	5	0
115	300	8	0
149	300	10	0
178	300	15	0
229	300	20	0
5	100	2	3.9
43	100	5	3.9
76	100	10	3.9
4	100	2	19.5
57	100	5	19.5
83	100	10	19.5
6	100	2	39.0
57	100	5	39.0
84	100	10	39.0

Data courtesy of Drs Morse, McKinlay and Spurr of Agriculture Canada.

10.5.2 *Assay of an insecticide with a synergist*

The data for this example are shown in Table 10.2 and come from a forthcoming paper by Morse, McKinlay and Spurr on the estimation of lowest-cost mixtures of insecticides and synergists. They relate to assays on a grasshopper *melanoplus sanguinipes* (F.) with the insecticide carbofuran and the synergist piperonyl butoxide (PB) which enhances the toxicity of the insecticide. The first model to be tried is of a type suggested by Hewlett (1969) and having the form of a logit link and binomial error with 'linear' predictor given by

$$\eta = \alpha + \beta_1 x_1 + \frac{\beta_2 x_2}{\phi + x_2},$$

where x_1 is the log dose of insecticide and x_2 is the dose of the synergist PB. The effect of the synergist is thus modelled as affecting the intercept by adding a hyperbolic term tending to β_2 for large x_2. The slope of β_1 is assumed unaffected by the amount of PB. If ϕ were known we could set $u = x_2/(\phi + x_2)$ and a generalized linear model would result. To estimate ϕ we set up u for some starting value of

ϕ and include the derivative $\partial u/\partial\phi = -u^2/x_2$ as a further covariate. Starting with $\phi = 1$ the standard process converges in four iterations to $\hat{\phi} = 1.763$ and a deviance of 53.34 with 11 d.f. The fit is poor with a deviance nearly five times the base level of 11.

Inspection of the residuals shows that the major part of the discrepancy comes from the low doses of insecticide where the fitted kills are all considerably greater than those measured. The alternative links, probit and complementary log–log, give very similar results, suggesting that the log dose is not a satisfactory scale for the insecticide. The low kills for the low doses suggest that there may be a threshold value for the insecticide, and we can test this by putting a second non-linear parameter θ in the model to represent the threshold. The model now takes the form

$$\eta = \alpha + \beta_1 \ln(z-\theta) + \beta_2 x_2/(\phi + x_2),$$

where z is the dose of insecticide. Given current values, θ_0 and ϕ_0, of θ and ϕ, the linearized form is given by

$$\eta = \alpha + \beta_1 \ln(z-\theta_0) - \gamma_1\left(\frac{1}{z-\theta_0}\right) + \beta_2\left(\frac{x_2}{\phi + x_2}\right) - \gamma_2 \frac{x_2}{(\phi + x_2)^2},$$

and with starting values $\phi_0 = 1.76$ from the first model and $\theta_0 = 1.5$ the estimation process again converges quickly to give estimates $\hat{\theta} = 1.67$, $\hat{\phi} = 2.06$, with a deviance of 18.70 with 10 d.f., clearly a great improvement on the first model. A test of variation in the slope β_1 with level of x_2 now gives no significant reduction in the deviance, whereas with the first model the deviance was nearly halved by allowing the slope to vary. A final iteration multiplying the two derivative covariates by b_1 and b_2 and using the mean deviance as a heterogeneity factor gives the estimates, standard errors and correlations shown in Table 10.3.

Table 10.3 *Results of the analysis of the insecticide–synergist assay*

Parameter	*Estimate*	*SE*	*correlations*			
α	−2.896	0.340				
β_1	1.345	0.143	−0.97			
θ	1.674	0.154	0.78	−0.77		
β_2	1.708	0.241	−0.31	0.26	−0.03	
ϕ	2.061	1.49	0.09	−0.06	0.07	0.61

Note that the two non-linear parameters are estimated almost independently (correlation −0.06), and that ϕ is ill-determined. In particular ϕ must be positive, so that the standard error is suspect; a plot of the deviance against fixed values of ϕ near 2 shows a curve that is non-quadratic. The use of a square-root transformation for ϕ is a great improvement and indicates that confidence limits for ϕ should be calculated on the square-root scale. The fitting of a separate intercept for each level of synergist in place of the term in $x_2/(\phi+x_2)$ gives a trivial reduction in deviance, indicating that the hyperbolic form is satisfactory. There remain two large residuals, for units 1 and 2, of opposite sign, whose removal from the fit reduces the deviance from 18.70 to 5.69; however, there seems no good reason to reject them. The relevant values are:

Unit	r	n	$\hat{\mu}$	r_P
1	7	100	14.67	−2.17
2	59	200	43.51	2.66

10.5.3 *Mixtures of drugs*

Darby and Ellis (1976) quote the data of Table 10.4, where the response y is the conversion of (3-3H)glucose to toluene-extractable lipids in isolated rat fat cells, and the two drugs are insulin in two forms, (1) standard and (2) A1-B29 suberoyl insulin. These are given in seven different mixtures, each at two total doses; there are four replicate readings for the 14 treatments. Darby and Ellis propose the model

$$\mathrm{E}(Y_{ijk}) = \alpha+\beta \ln(x_{1ij}+\theta x_{2ij}) \tag{10.2}$$

with constant-variance errors. Here i indexes the mixtures, j the degree dose and k the replicates, while x_{1ij} and x_{2ij} are the amounts of the two drugs given for mixture i with dose j.

Here θ is the non-linear parameter and we can fit the model by linearizing it, using the two covariates

$$u = \ln(x_1+\theta x_2), \qquad v = \frac{\partial u}{\partial \theta} = \frac{x_2}{x_1+\theta x_2},$$

and fitting $\alpha+\beta u+\gamma v$ for some value θ_0; θ_1 is then updated by $\theta_1 = \theta_0+\gamma/\beta$. The estimate obtained is $\hat{\theta} = 0.0461 \pm 0.0036$, with a corresponding deviance (residual sum of squares) of 244.0 with 53 d.f.

Table 10.4 *Results of insulin assay*

Mixture	*Ratio of insulin to A1-B29 suberoyl insulin*	*Total dose* (pmol l^{-1})	*Responses for four replicates*			
1	1:0	20.9	14.0	14.4	14.3	15.2
		41.9	24.6	22.4	22.4	26.7
2	1:1.85	52.9	11.7	15.0	12.9	8.3
		106	20.6	18.0	19.6	20.5
3	1:5.56	101	10.6	13.9	11.5	15.5
		202	23.4	19.6	20.0	17.8
4	1:16.7	181	13.8	12.6	12.3	14.0
		362	15.8	17.4	18.0	17.0
5	1:50.0	261	8.5	9.0	13.4	13.5
		522	20.6	17.5	17.9	16.8
6	1:150	309	12.7	9.5	12.1	8.9
		617	18.6	20.0	19.0	21.1
7	0:1	340	12.3	15.0	10.1	8.8
		681	20.9	17.1	17.2	17.4

Data from Darby and Ellis (1976).

Comparing this fit with the replicate error of 154.8 with 42 d.f., we find an F-value for residual treatment variation of $F(11, 42) = 2.20$, just beyond the 5% point. Darby and Ellis are concerned to compare this model with one in which θ is allowed to vary with the mixture, and in doing so to provide possible evidence of synergism or antagonism. Such a model, which requires a set of partial-derivative covariates, one for each mixture, reduces the deviance to 194.6 with 48 d.f., giving a residual $F(6, 42) = 1.80$. While no longer significant at 5%, there is still a noticeable lack of fit, which investigation shows to be entirely concentrated in the first 'mixture', that for which x_2 is zero. Without this we find $\hat{\theta} = 0.0524$ and a deviance of 191.2 with 51 d.f., giving a residual treatment deviance of 36.40 with 9 d.f., so that the mean deviance is now close to the replicate error mean square of 3.686. On this interpretation the interaction between the two drugs is expressed by saying that one unit of drug 2 is equivalent to 0.052 units of drug 1 in mixtures containing ratios by weight of more than 1:1.85. In the absence of drug 2 the actual response to drug 1 is larger than predicted from the model (10.2), the predicted

Table 10.5 *Analysis of variance for insulin assay*

	s.s.	*d.f.*	*m.s.*
Treatments	906.6	13	
Model (10.2)	817.4	2	408.7
Separate θs	49.4	5	9.88
Residual	39.8	6	6.63
Alternative subdivision			
Model (10.2)	817.4	2	
Removal of mixture 1	52.8	2	26.4
Residual	36.4	9	4.04
Within treatments	154.8	42	3.686

values for 'mixture' 1 (omitting it from the fit) being 12.9 and 19.8 as against 14.5 and 24.0 actually measured. Plots of residuals from the restricted model show no pattern when plotted against fitted values or mixture number, and there are no obvious outliers. The final analysis of variance is given in Table 10.5; it should be noted that this analysis differs somewhat from that of Darby and Ellis.

CHAPTER 11

Model checking

11.1 Introduction

When a model is matched against data for the purposes of summarization and prediction it is obviously desirable that it should be selected after careful prior consideration of its likely relevance. Thus, for example, in modelling counts fitted values from a model should be confined to non-negative values, because counts are. Similarly if it is known *a priori* that a response to a stimulus variable x tails off beyond a certain level which is well within the range of x in the data, then a linear term only in x will not be adequate for the model.

Prior information, however, will not always be enough, and we need to supplement the fitting of a model with a subsequent inspection of that fit. This chapter deals with the checking of generalized linear models; we extend, where possible, techniques originally devised for regression models, and develop new ones for aspects of the general class which have no analogues in the restricted one.

11.2 Techniques in model checking

Methods of model checking fall into three main categories:

1. tests of deviations in particular directions,
2. visual displays,
3. the detection of influential points.

Category-1 methods are implemented by including extra parameters in the model, and testing whether their inclusion significantly improves the fit or not. Extra parameters might arise from including an extra covariate, from embedding a covariate x in a family $f(x;\theta)$ indexed by an extra parameter θ, from embedding a link function $g(\eta)$ into a similar family $g(\eta;\theta)$, or from including a constructed variate, say $\hat{\eta}^2$, depending on the original fit. Category-1 methods thus look for deviations in the fit in certain definite directions thought likely to be important *a priori*.

By contrast, category-2 methods rely mainly on the human mind and eye to detect pattern. Such methods take a successful model to be one that leaves a patternless set of residuals (or other derived quantities). The argument is that if we can detect pattern we can find a better model; the practical problem is that any finite set of residuals can be made to yield some kind of pattern if we look hard enough, so that we have to guard against over-interpretation. Nonetheless category-2 methods are an important component in model checking.

Category-3 methods, when applied to Normal linear models, have become known as regression diagnostics. Their aim is the detection of data points that are in some sense particularly influential in the fitting; this may imply (but does not necessarily do so) that the inferences made from the model are especially heavily dependent on certain dependent response values or covariate values. It is important that the analyst should be aware of such points and their influence on his inferences. Influential points will often be outliers, in the sense of lying some distance from the main mass of data values; such outliers need to be identified so that they can be carefully checked in the original data, because they may, of course, be simply wrong values, the result of transcription or recording errors.

11.3 The raw materials of model checking

In regression models, model checking is based mainly on certain quantities derived from the data matrix, namely the fitted-value vector $\hat{\boldsymbol{\mu}}$, the residual variance s^2 and the diagonal elements of the projection matrix, $\mathbf{H} = \mathbf{X}(\mathbf{X}^T\mathbf{X})^{-1}\mathbf{X}^T$, which maps $\mathbf{y}$ into $\hat{\boldsymbol{\mu}}$. In generalized linear models we may use the linear-predictor vector $\hat{\boldsymbol{\eta}}$ or other functions of $\hat{\boldsymbol{\mu}}$ in place of $\hat{\boldsymbol{\mu}}$, and the adjusted dependent variate $\mathbf{z}$ in place of $\mathbf{y}$. The $\mathbf{H}$ matrix becomes $\mathbf{W}^{\frac{1}{2}}\mathbf{X}(\mathbf{X}^T\mathbf{W}\mathbf{X})^{-1}\mathbf{X}^T\mathbf{W}^{\frac{1}{2}}$.

A central role is played by the residuals from the fit; three forms of theoretical residual were defined in Section 2.4, the Pearson residual, the Anscombe residual and the deviance residual. They were defined in terms of μ rather than $\hat{\mu}$; the estimation of μ in the actual residual requires a standardization factor, so that the Pearson residual

$$r_P = (y - \hat{\mu})/\sqrt{\text{var}(Y)}$$

is replaced by the standardized Pearson residual

$$_{S}r_{P} = (y-\hat{\mu})/\sqrt{\mathrm{var}(Y-\hat{\mu})}. \tag{11.1}$$

Here the denominator is $\sqrt{[V(\hat{\mu})(1-h)]}$ where $V(\ )$ is the variance function and h is the diagonal element of $\mathbf{H}$. The standardizing factor $\sqrt{(1-h)}$ becomes important if the variances of $\hat{\mu}$ vary widely between points or if the number of parameters is an appreciable fraction of N. The calculations of Cox and Snell (1968) support a similar standardization for the deviance residual r_D. It should be noted, however, that whereas for linear regression models with Normal errors standardization removes terms of order N^{-1} from the variance, making it exactly σ^2, the same is not true for residuals from non-linear generalized linear models, so that terms of order N^{-1} remain in both the mean and variance. However, this does not seem of great importance in practice.

11.4 Tests of deviations in particular directions

Having fitted a generalized linear model, specified by its variance function, link function and covariates in the linear predictor, we may then ask questions such as:

1. Is a further covariate required in the linear predictor?
2. Should a given covariate x, say, be replaced by a function, say $\ln x$?
3. Should the link function be changed in a certain direction, e.g. from $\eta = \mu$ to $\eta = \ln \mu$?
4. Should the variance function be changed from, say, $V(\mu) = \lambda\mu$ to $V(\mu) = \lambda\mu^2$?

Question 1, about which there exists an enormous literature, was briefly discussed in Section 3.9, and will not be further considered here. The extension of the ideas from regression models to generalized linear models does not pose any basically new problems.

11.4.1 *Checking a covariate scale*

The usual technique involves embedding the scale in a family, e.g. by replacing x by x^θ (with $\ln x$ for $\theta = 0$), and then estimating θ by the methods discussed in Section 10.4. The change in deviance is then assessed when the prior value (here $\theta = 1$) is changed to the value of best fit, $\theta = \hat{\theta}$. A graphical category 2 technique is described in Section 11.5.1.

11.4.2 *Checking the link function*

Again embedding may be employed, with the prior link $\eta = g(\mu)$ expanded to $\eta = g(\mu;\theta)$ and the deviance compared as before for $\theta = \theta_0$ and $\theta = \hat{\theta}$. Fitting techniques are described in Section 10.3. A useful indicator of the adequacy of a link function is the change in deviance after including the additional covariate $\hat{\eta}^2$ in the linear predictor, following the original fit.

For the car insurance data inclusion of $\hat{\eta}^2$ as an extra covariate after fitting main effects on the inverse scale with gamma errors produces a trivial reduction in the deviance from 124.7 with 109 d.f. to 124.6 with 108 d.f. However, with an identity link the inclusion of $\hat{\eta}^2$ reduces the deviance from 139.7 to 126.2, pointing to an unsatisfactory link. A graphical method is described in Section 11.5.

11.4.3 *Checking the variance function*

The problem with comparing variance functions is that the deviance is no longer a suitable criterion. Recent work by Nelder and Pregibon (1983) suggests a possible criterion for positive continuous data, its form for discrete data being less clear, however.

For the six distributions used in generalized linear models, the Normal, Poisson, binomial, negative binomial, gamma and inverse Gaussian, it is found that the log likelihood l is close to an extended quasi-likelihood l' defined by

$$l' = \frac{1}{\phi}(Q - Q_0) - \tfrac{1}{2}\ln[2\pi\phi V(y)], \qquad (11.2)$$

where Q is the quasi-likelihood kernel for a variance function $V(\mu)$, i.e.

$$Q = \int \frac{y-\mu}{V(\mu)}\,\mathrm{d}\mu,$$

Q_0 is the same kernel for $\mu = y$, ϕ is the dispersion parameter, and $V(y)$ is the variance function evaluated at the datum value y. For the Normal and inverse Gaussian distributions l is exactly equal to l'; for the gamma distribution l differs from l' by a factor depending on ϕ; and for it and for the remaining discrete distributions l' is obtained by replacing all factorials in l by the Stirling approximation

$$k! \simeq \sqrt{(2\pi k)} k^k e^{-k}.$$

Thus l' is the saddle-point approximation for exponential families discussed by Barndorff-Nielsen and Cox (1979) but applied here to single observations. Note, however, that l' is also defined for quasi-likelihood models where only the mean and variance function are assumed.

Quasi-likelihood l' has the following properties:

1. For a generalized linear model with $\mu = g(\Sigma \beta_j x_j)$ the estimates of $\boldsymbol{\beta}$ obtained from maximizing l' are the maximum-quasi-likelihood estimates obtained from maximizing Q, and thus are the maximum-likelihood estimates when a distribution exists.
2. The estimate of the dispersion parameter ϕ from maximizing l' is given by equating $\hat{\phi}$ to the mean deviance. This is the same as the maximum-likelihood estimate for the Normal and inverse Gaussian distributions, because for these $l = l'$.
3. The maximum value of l' after estimating $\boldsymbol{\beta}$ and ϕ is given by

$$-2l'_{\max}(\hat{\boldsymbol{\beta}}, \hat{\phi}) = N + \Sigma \ln[2\pi\hat{\phi} V(y)].$$

The log-likelihood-ratio statistic for two variance functions $V_1(\mu)$ and $V_2(\mu)$ with corresponding deviances D_1 and D_2, given the same link and linear predictor, is thus given by

$$Z = N \ln(D_1/D_2) + \Sigma \ln[V_1(y)/V_2(y)].$$

Table 11.1 shows the values of Z for a family of variance functions of the form $V(\mu) = \phi\mu^\theta$ for the car insurance data (the example in Section 7.4.1), using the inverse link. Models were fitted for $\theta = 0, 1, 2, 3$ and the results show that the initial assumption of $\theta = 2$ was satisfactory.

Table 11.1 *The criterion Z for the family of variance functions $V = \mu^\theta$ applied to the car insurance data*

θ	Z
0	41.1
1	13.8
2	0
3	9.5

For discrete distributions the use of the Stirling approximation above is unsatisfactory because it precludes the value zero. This deficiency may be removed by adopting instead the approximation

$$k! \simeq \sqrt{[2\pi(k+c)]}k^k e^{-k} \tag{11.3}$$

where $c = 1/6$. For large k this introduces the term $1/12k$ into the expansion of $\ln k!$, as required in the Stirling asymptotic series, while giving the limiting value 1.023 for 0!. For $k \geqslant 1$ the approximation has a relative error of less than 0.4%. Use of (11.3) leads to a form of l' in which $V(y)$ is replaced by a modified function $V'(y)$, for which the three distributions give the following forms

Poisson	$y+c$,
binomial	$\dfrac{(np+c)(nq+c)}{n+c}$,
negative binomial	$\dfrac{(y+\nu)^2(y+c)(\nu+c)}{\nu^2(y+\nu+c)}$.

For ν not too small, $V'(y)$ for the negative binomial is close to $(y+c)(y+\nu+c)/\nu$, i.e. the expression obtained by substituting $y+c$ for μ in the variance function $\mu(\mu+\nu)/\nu$. We are thus led to the following modified form of l' for some discrete quasi-likelihood models: for counts with a variance function of the form $\lambda\mu+\theta\mu^2$ use $V'(y) = \lambda(y+c)+\theta(y+c)^2$ and for proportions use $V'(y)$ obtained from $V(\mu)$ by substituting $(np+c)/(n+c)$ for p, $(nq+c)/(n+c)$ for q and $n+c$ for n. Thus for Wedderburn's variance function np^2q^2 (see the example in Section 8.6.1) we would use

$$V'(y) = (np+c)^2(nq+c)^2/(n+c)^3.$$

Replacement of $V(y)$ by $V'(y)$ leaves the estimates of $\boldsymbol{\beta}$ and ϕ unchanged, while allowing the statistic Z to be defined for discrete data when comparison of variance functions is required.

Graphical methods for checking the variance function are described in Section 11.5.3.

11.4.4 *Score tests for extra parameters*

Many procedures used in model checking can be shown to be special cases of the class of score tests (Rao, 1973, ch. 6). Consider two models, one (M_0) with p parameters and a second (extended) model (M_1) with $p+k$ parameters. The deviance test involves the fitting of

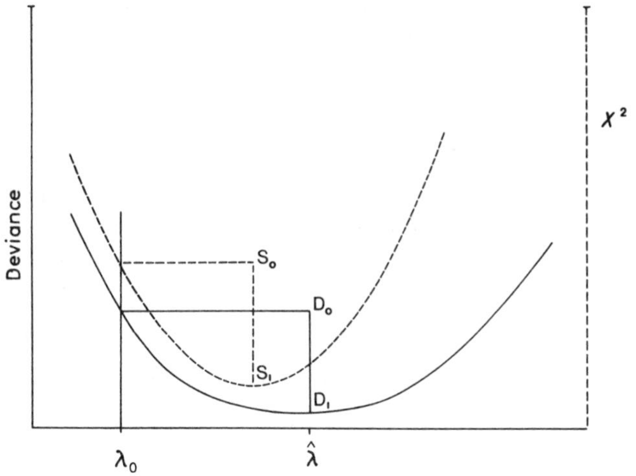

Fig. 11.1. *The geometry of the score test: dotted line, the curve of X^2 using adjusted dependent variate and weights from fit at $\lambda = \lambda_0$; solid line, the deviance for varying λ.*

both M_0 and M_1 and a test of the difference of the associated deviances. For generalized linear models the score test involves the fitting of M_0 followed by the fitting of M_1 as linear predictor but using the quadratic weight function and adjusted dependent variate appropriate to M_0; the score-test statistic is then the difference in X^2, rather than in deviance, between the two fits. Pregibon (1982) gives the details. The computing advantage of the score test over the deviance test is that it involves only a single iteration for the more complex model, compared with iteration to convergence for the deviance (or likelihood-ratio) test. Note that the two tests give identical results for linear models. Their relationship for one extra parameter is shown in Fig. 11.1. The solid curve is of the minimum deviance for the model M_1 for varying values of the extra parameter λ, M_0 itself assuming a fixed value λ_0. The deviance test measures the difference in ordinates D_0D_1, while the score test uses the first and second derivatives of the log likelihood at λ_0 to predict the position of the minimum, assuming a quadratic shape for the curve. The score-test statistic thus measures the difference S_0S_1. Clearly the agreement of the two tests for generalized linear models will depend on how well the solid curve approximates a parabola between

λ_0 and $\hat{\lambda}$. A check on this could be made by iterating the fit of the extended model M_1 through a second cycle; a marked change in X^2 would then indicate lack of quadraticity in the surface.

11.5 Visual displays

An important plot after fitting is that of residuals against fitted values (with the latter perhaps transformed). Such a plot is capable of revealing isolated points with large residuals, or a general curvature, indicating unsatisfactory covariate scales or link function, or a trend in the spread with increasing fitted values, indicating an unsatisfactory variance function.

With Normal models we plot $(y-\hat{\mu})$ (perhaps standardized) against $\hat{\mu}$, and on this plot the contours of constant y are 45° lines running from NW to SE of the diagram. We can preserve this property at the point $r = 0$ for generalized linear models by choosing a scale for $\hat{\mu}$ for which the slopes of the contours of $y =$ constant are -1 at $r = 0$. This scale turns out to be one for which there is constant information, i.e. it is given by $\int d\mu/V^{\frac{1}{2}}(\mu)$. Thus for Poisson errors we use $2\sqrt{\hat{\mu}}$ for the horizontal scale, for binomial errors $2\sin^{-1}(\sqrt{\hat{\mu}})$ and for gamma errors $\ln \hat{\mu}$. Other choices for the horizontal scale are of course possible, including $\hat{\mu}$ and $\hat{\eta}$. Fig. 11.2 shows the deviance residual plotted against $\ln \hat{\mu}$ for the car insurance data, after fitting main effects on the inverse scale with gamma errors. The appearance of the plot is satisfactory, with no obvious curvature or systematic change of range of residuals along the horizontal scale.

The residual–fitted-value plot is inevitably distorted with discrete data in that extreme values of the data give finite limits for any residual. Thus for Poisson data the Pearson residual is bounded below by $-\hat{\mu}^{\frac{1}{2}}$, the Anscombe residual by $-1.5\hat{\mu}^{\frac{1}{2}}$ and the deviance residual by $-\sqrt{(2)}\hat{\mu}^{\frac{1}{2}}$. Moreover the distribution of the residuals for small μ is markedly skew, making the plot relatively uninformative in the region $\hat{\mu} < 2$. Similar considerations apply to proportions with fitted values near either zero or unity.

A commonly used technique to assess extreme residuals is to order the standardized residuals and plot them against the expected Normal order statistics. Extreme residuals then appear at the ends and, if unduly large, deviate from the main trend. This procedure makes allowance for selecting the largest residuals for inspection but

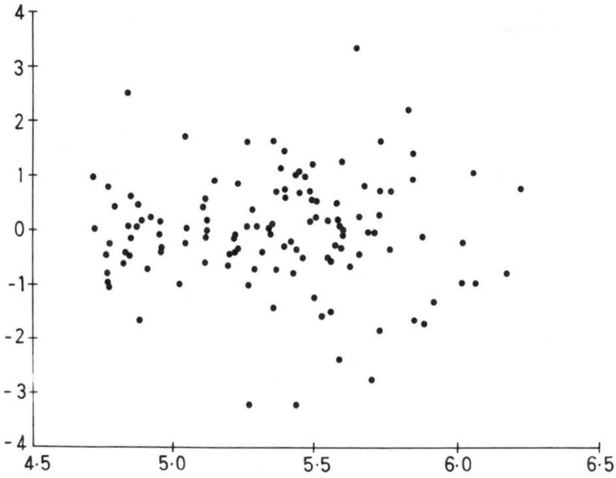

Fig. 11.2. *Car insurance data: deviance residuals vs ln(fitted values) for model with gamma errors and main effects with reciprocal link.*

depends upon an assumption that the residuals are, at least approximately, Normally distributed. With discrete data having many zeros and associated small values of $\hat{\mu}$, the bounds on residuals may cause an apparent discontinuity in the middle of this plot, caused by the clustering of small negative residuals.

We now consider visual analogues of the formal tests of Section 11.4.

11.5.1 *Checking form of covariate*

A plot due to Pregibon (personal communication) uses the *partial residual* for the covariate in question, x say, expressed on the scale of the adjusted dependent variable z. Let $\hat{\eta}$ be the linear predictor for a model including x, then we define the generalized partial residual by

$$u = z - \hat{\eta} + \hat{\gamma}x,$$

where $\hat{\gamma}$ is the regression coefficient for x. We then plot u against x; this plot should be approximately linear if the scale of x is correct; if not its form may suggest a suitable alternative. Note that the variance of U will not be constant.

11.5.2 *Checking form of link function*

For this a similar form of argument suggests that we plot $z-\hat{\eta}+\hat{\eta}$, i.e. z, against $\hat{\eta}$. Again the form of the plot, if not linear, will suggest an appropriate transform, and the variance of Z will not be constant.

11.5.3 *Checking the variance function*

The plot of residuals against fitted values can be used to check if the spread of residuals is approximately constant and independent of $\hat{\mu}$. Note that intervals on the $\hat{\mu}$ scale with a high density of points will, other things being equal, produce a wider range of residuals than an interval with a low density, and some attempt must be made to allow for this in scanning the plot. A variance function which increases too/insufficiently rapidly with μ will cause the scatter of residuals to decrease/increase with increasing μ. It may be difficult, indeed impossible, to distinguish the presence of a few outliers from the use of the wrong variance function, and many procedures for outlier detection are sensitive to the assumption of the form of the variance function. With discrete data the portion of the plot near the limiting values of $\hat{\mu}$ again cannot be used.

An alternative plot uses $|r|$ in place of r. This puts large residuals of either sign at the top, and so concentrates the information on one side of the plot. Fig. 11.3 illustrates the two plots for the car insurance data. The residuals are deviance residuals, scaled by $s\sqrt{(1-h)}$, where s^2 is the estimated dispersion parameter. One plot is shown for a variance function that increases too slowly, namely $V(\mu) = \lambda\mu$, which clearly shows the increasing trend with increasing fitted value; the other is shown for a variance function that increases too rapidly, namely $V(\mu) = \lambda\mu^3$, which shows a corresponding decreasing trend.

11.6 Detection of influential observations

At the opposite extreme from category-1 procedures which deal with deviations detectable in the pattern of the data as a whole, we have category-3 procedures which are concerned to isolate individual (or small sets of) observations having most influence on the fit. It is well known that residuals as such cannot detect influential points, because (a) points that deviate most from the pattern suggested by the rest

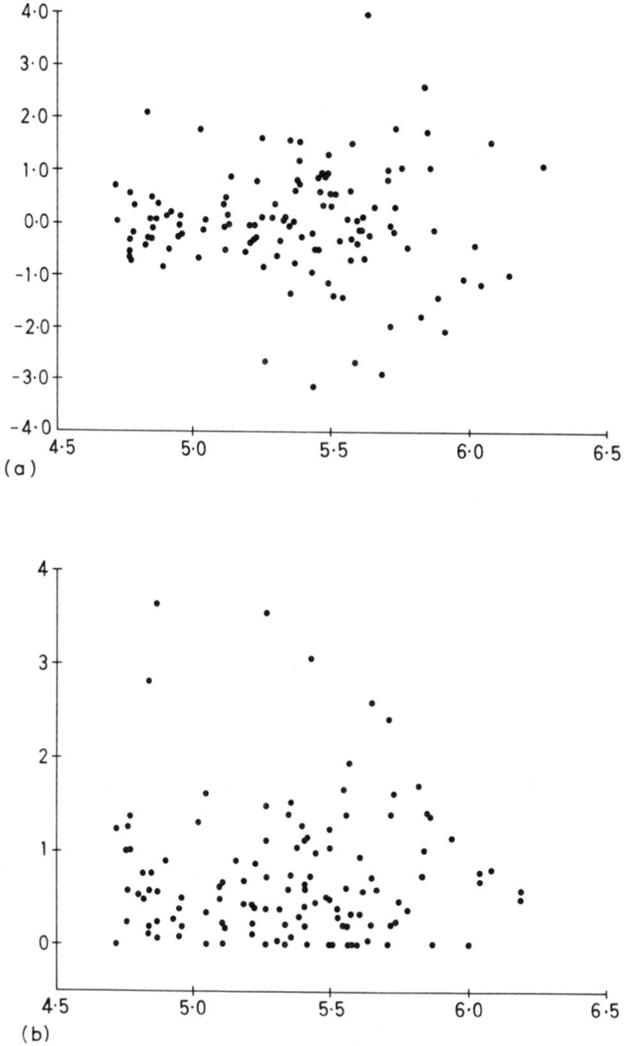

Fig. 11.3. *Car insurance data: two plots showing unsatisfactory variance functions; systematic part of model – main effects with reciprocal link.* (*a*) *Residuals vs ln(fitted values) for* $V(\mu) = \lambda\mu$, *and* (*b*) |*residuals*| *vs ln(fitted values) for* $V(\mu) = \lambda\mu^3$.

need not have large residuals, and (b) points with large residuals may affect the fit very little.

The important quantities for detecting points of high influence are the diagonal elements h_i of $\mathbf{H} = \mathbf{W}^{\frac{1}{2}}\mathbf{X}(\mathbf{X}^{\mathrm{T}}\mathbf{W}\mathbf{X})^{-1}\mathbf{X}^{\mathrm{T}}\mathbf{W}^{\frac{1}{2}}$. This projection matrix is the ratio of the covariance matrix of $\hat{\boldsymbol{\mu}}$ to the covariance matrix of $\mathbf{Y}$. For points of high influence, h_i will be near unity. The average value of h_i is p/N, where p is the number of parameters and N the number of observations. Hoaglin and Welsch (1978) suggest using $h_i > 2p/N$ as a rule of thumb to indicate influential points. By definition the standardized residuals (11.1) will not reveal such points; Cook (1977) suggested a statistic, which we generalize in the form

$$D_i = (\hat{\boldsymbol{\beta}}_{(i)} - \hat{\boldsymbol{\beta}})^{\mathrm{T}} (\mathbf{X}^{\mathrm{T}}\mathbf{W}\mathbf{X}) (\hat{\boldsymbol{\beta}}_{(i)} - \hat{\boldsymbol{\beta}})/ps^2$$

as a measure of the effect of the ith datum point on the parameter estimates, where $\hat{\boldsymbol{\beta}}_{(i)}$ denotes the estimates obtained when that point is omitted, and s^2 is an estimate of the dispersion parameter.

An analogous form of residual is the jack-knife or cross-validatory residual, which is based on standardizing $y - \hat{\mu}_{(i)}$ with $s_{(i)}$, where again a suffix in brackets denotes estimates obtained when point i is omitted. Although the cross-validatory residual is a monotone function of the ordinary standardized residual, it reflects large deviations more dramatically by tending to infinity as an upper limit compared with the maximum of $\sqrt{(N-p)}$ for the ordinary residual.

Atkinson (1981a) proposes the use of the cross-validatory residuals r_i^* and a set of quantities T_i derived from Cook's statistic for the detection of influential y and x values respectively. T_i is derived from D_i by replacing r_i by r_i^*, using the square-root scale, and standardizing for the effect of N against a D-optimum design. The result is given by

$$T_i = \left[\left(\frac{N-p}{p}\right)\left(\frac{h_i}{1-h_i}\right)\right]^{\frac{1}{2}} \mid r_i^* \mid .$$

To aid the interpretation of the half-normal plots of r^* and T, Atkinson constructs an envelope by simulating 19 additional sets of pseudo-random Normal variables, fitting the same model to them and plotting the maximum and minimum values of each ordered statistic across the 19 sets. For generalized linear models this process would require the construction of pseudo-data sets based on the same $\hat{\mu}$ with the correct variance function, followed by the fitting of the generalized linear model, construction of r^*, etc. More work is required on the properties of such generalized envelopes.

A problem with all the category-3 methods discussed is that they deal essentially with the effect of a single influential observation. If several observations are influential, then the plots reflecting one-at-a-time deletion may be relatively uninformative about the effect of simultaneous deletion of more than one. The combinatorial explosion causes difficulties in looking at all possible sets of two, three, ... observations, and there is an obvious added problem of over-interpretation.

11.7 Bibliographic notes

Recent books on the subject of model checking include Belsley *et al.* (1980), Hawkins (1980) and Cook and Weisberg (1982). These authors emphasize Normal-theory linear models although Cook and Weisberg do include a section on generalized linear models. Cox and Snell (1968) discuss residuals in a very general context including topics such as standardization and non-linear transformations. Goodness-of-link tests are discussed by Pregibon (1980) and by Atkinson (1982). The technique of using a constructed variable for detecting departures of a specific nature goes back to Tukey (1949). Other useful references are Andrews and Pregibon (1978) and Pregibon (1979). Cook (1977, 1979) introduced the notion of influential observations in regression. Atkinson (1981a) proposed a modification of Cook's statistic for the detection of influential x-values.

A more extensive bibliography can be found in the three books mentioned above.

CHAPTER 12

Further topics

In this chapter we give a very brief account of selected areas of current research and indicate important topics requiring further work.

12.1 Second-moment assumptions

In Chapter 8 and Appendix C we discuss the notion of quasi-likelihoods which are based on second-moment assumptions. From an applied viewpoint the use of second-moment assumptions is attractive because (a) there is the possibility of making a check on the assumptions and (b) it would be unwise to use a method that relies heavily on an assumption that is possibly false and not easy to verify. However, the restriction imposed in Chapter 8 that $\text{cov}(\mathbf{Y}) = \sigma^2 V(\boldsymbol{\mu})$ is unduly restrictive and excludes, for example, autoregressive models. It would be interesting and useful to develop methods capable of dealing with the case where $\text{cov}(\mathbf{Y}) = \mathbf{V}(\boldsymbol{\mu}, \boldsymbol{\rho})$ where, for example, $\boldsymbol{\rho}$ might be a vector of autocorrelations.

Secondly, all other things being equal, efficient estimates are preferable to inefficient ones. However, if increased efficiency is achieved at a substantial cost in terms of reduced robustness, the comparison is more difficult. It would be useful to have a comparison of the efficiencies and robustness of quasi-likelihood estimates with likelihood estimates in some cases where the two are different, as in the example of Section 8.6.2.

12.2 Higher-order asymptotic approximations

In small- or medium-sized samples the approximations described in Appendix C for quasi-likelihoods may be unsatisfactory. To obtain a more refined approximation it is usually necessary to make stronger assumptions such as, say, the nature of the third and fourth moments. If, further, the distribution is completely specified, it may be possible to obtain an exact theory: see, for example, Sections 4.4 and 5.4.

In the absence of an exact small-sample theory it may be possible to obtain the moments of $\hat{\boldsymbol{\beta}}$ to a higher order of approximation (Cox and Snell, 1968; Barndorff-Nielsen, 1980; Durbin, 1980). Also the distribution of the likelihood ratio may be found to a higher order of approximation (Lawley, 1956; Williams, 1976; Cordeiro, 1983a, b). In very simple cases the asymptotic calculations given in Appendices A, B and D may be used.

For scalar parameters a one-sided test is usually required although the likelihood-ratio statistic gives a two-sided test. However, if we take the signed square root of the likelihood-ratio statistic

$$W^{\frac{1}{2}}(\beta_0) = \pm[2l(\hat{\beta}; y) - 2l(\beta_0; y)]^{\frac{1}{2}},$$

where the sign of $W^{\frac{1}{2}}(\beta_0)$ is that of $\partial l/\partial\beta_0$, we may use the approximation

$$W^{\frac{1}{2}}(\beta_0) + \rho_3/6 \sim \mathrm{N}(0, 1). \qquad (12.1)$$

Here ρ_3 is the third standardized cumulant of $\partial l(\beta; Y)/\partial\beta$ at β_0 and is of the order $N^{-\frac{1}{2}}$.

There is a close similarity between the statistic $W^{\frac{1}{2}}$ and deviance residuals (Pregibon, 1979, 1981). Expression (12.1) suggests that deviance residuals should be corrected for bias.

When nuisance parameters are present the problem becomes considerably more complicated. Nevertheless Lawley's results continue to apply.

12.3 Composite link functions

In the standard model discussed in Chapter 3 the ith component of $\boldsymbol{\eta}$ depends on the ith component only of the mean vector $\boldsymbol{\mu}$ by the relation

$$\eta_i = g(\mu_i); \qquad i = 1, \ldots, N.$$

More generally, however, μ_i may depend on several components of $\boldsymbol{\eta}$. This extra level of generality raises no new theoretical problems: the treatment of Chapter 8 and Appendix C is sufficiently general to cover composite link functions which may be written

$$\boldsymbol{\mu} = \mathbf{g}^{-1}(\boldsymbol{\eta}).$$

One example is discussed in detail in Section 12.5, where the μs are multinomial cell means. There several cell means were added before the logistic transform was applied.

Thompson and Baker (1981) discuss composite link functions in general with particular emphasis on computing aspects. They give a number of additional examples of their application, including fused cells in a contingency table and grouped Normal data.

12.4 Missing values and the EM algorithm

Dempster *et al.* (1977) describe a general method for computing maximum-likelihood estimates when observations are missing. The essence of their EM algorithm is to begin with a starting value, $\hat{\boldsymbol{\beta}}_0$ say, for the parameters and to replace the complete-data log-likelihood l with its expectation given the observed data and $\hat{\boldsymbol{\beta}}_0$. This completes the E-step. The M-step involves the maximization of the resulting complete-data likelihood to obtain $\hat{\boldsymbol{\beta}}_1$, and the procedure is repeated until convergence occurs.

When the sufficient statistics of the complete data are linear functions of the data, the statement and execution of the algorithm may be simplified. In the E-step we replace the missing observations by their conditional expectations given the data observed and the parameter vector. In the M-step no distinction is made between the observed data and the imputed data and a new value $\hat{\boldsymbol{\beta}}_1$ is obtained by maximizing in the usual way.

The following points are relevant for implementing the EM algorithm in particular and for dealing with missing values in general.

1. The rate of convergence of the EM algorithm is often exceedingly slow: see the discussion of Dempster *et al.* (1977).
2. Unlike the method of weighted least squares, the EM algorithm does not yield variance estimates for the parameters.
3. If an observation has a covariate (as opposed to a response) missing, further assumptions are required to extract information from that particular observation. The simplest scheme replaces the missing x by its linear regression on the remaining xs. However, it is not clear in general whether there is any relationship between this procedure and maximum-likelihood estimation.

If the data including the covariates are jointly distributed according to the $(p+1)$-variate Normal distribution then the EM algorithm may be used as a recipe for estimating the parameters of this distribution by maximum likelihood. Because of the symmetry implied by the multivariate Normal density it makes little difference whether an x or a y is missing. However, in the context of logistic regression with missing xs the position is unclear. There is no 'natural' multivariate density that could be used for replacing the missing xs. The procedure described above seems intuitively reasonable but it is different from the EM algorithm and probably does not correspond to maximum-likelihood estimation.

12.5 Components of dispersion

In the analysis of designed experiments, such as split-plot designs, Normal models occur in which more than one component of error is present. Observational data may also require models with several error components for their analysis. Generalized linear models have so far been described with a single error component only, but there is an obvious need to search for analogues of the multicomponent types of Normal model. The problem is a difficult one but some progress can be made for a simple nested structure where the two components σ_b^2 and σ^2 measure respectively variation between and variation within blocks. In the usual components-of-variance problem it is assumed that the variance of the response is the same for each block regardless of the value of the response to be expected in the block. Here we introduce block parameters $\alpha_1, \ldots, \alpha_k$ that are random variables and we suppose that, conditionally on $\alpha_1, \ldots, \alpha_k$,

$$\begin{aligned} E(Y_{ij}|\boldsymbol{\alpha}) &= \alpha_i, \\ \operatorname{var}(Y_{ij}|\boldsymbol{\alpha}) &= \sigma^2 V(\alpha_i), \\ \operatorname{cov}(Y_{ij}, Y_{il}|\boldsymbol{\alpha}) &= 0; \qquad j \neq l. \end{aligned} \tag{12.2}$$

In other words, the within-block variance need not be constant across blocks.

The between-block component may be modelled by

$$E(\alpha_i) = \mu, \quad \operatorname{var}(\alpha_i) = \sigma_b^2 V(\mu), \quad \operatorname{cov}(\alpha_i, \alpha_j) = 0; \quad i \neq j, \tag{12.3}$$

where we have assumed that the same variance function, $V(\ )$, applies in both cases.

We now make the further assumption that, conditionally on $\alpha_1, \ldots, \alpha_k$ and for known σ^2, the block totals are sufficient for $\alpha_1, \ldots, \alpha_k$.

Thus we may write

$$\text{var}(Y_{ij}|\bar{Y}_i) = \sigma^2\left(\frac{N_i-1}{N_i}\right)b(\bar{y}_i;\sigma^2, N_i)$$

independently of α_i, for some function $b(\ )$. Here N_i is the size of the ith block. Using the usual formula connecting conditional and unconditional variances and temporarily dropping subscripts, we have that

$$\sigma^2 V(\alpha) = \sigma^2\left(\frac{N-1}{N}\right)\text{E}(b(\bar{Y};\sigma^2, N)|\alpha)+\frac{\sigma^2 V(\alpha)}{N}.$$

It follows that the conditional variance function, $b(\ )$, is given in terms of $V(\ )$ by

$$b(\ ) = V(\ )[1-\sigma^2 V''/(2N)-N^{-\frac{3}{2}}\sigma^4 V'V'''/6+\text{O}(N^{-2})]. \quad (12.4)$$

The first neglected term is $[\sigma^2 V''/(2N)]^2$ in the special but important case where $V(\ )$ is quadratic in μ. In applications with moderate to large values of N, the approximation $\sigma^2 V(\ )[1-\sigma^2 V''/(2N)]$ should generally be adequate.

Let the sample means and sample variances be $y_1, ..., y_k$ and $s_1^2, ..., s_k^2$ respectively. The sample variances have conditional expectations

$$\text{E}(s_i^2|\bar{Y}_i) = \sigma^2 b(\bar{y}_i;\sigma^2, N_i); \qquad i = 1, ..., k.$$

Thus we may construct a pooled estimator, σ^2, of the within-blocks dispersion component. In other words we solve for s^2 the equation

$$(N.-k)s^2 = \Sigma(N_i-1)s_i^2/b(\bar{y}_i;s^2, N_i). \quad (12.5)$$

Equation (12.5) guarantees consistency conditionally and therefore unconditionally but it does not guarantee unbiasedness except in the special cases where $V'' = 0$. In this special case the expression on the right of (12.5) reduces to $\Sigma(N_i-1)s_i^2/V(\bar{y}_i)$, an obvious analogue of Pearson's statistic computed within blocks and pooled over blocks. More generally when $V'' \neq 0$ (12.5) must be solved iteratively. An obvious starting point is

$$(N.-k)^{-1}\Sigma(N_i-1)s_i^2/V(\bar{y}_i).$$

The between-blocks dispersion component can be estimated in a similar way. We deal for simplicity with the balanced case $N_1 = ... = N_k = N$. The approximate unconditional variance of $\bar{Y}_i$ is

$$[\sigma_b^2+\sigma^2(1+\tfrac{1}{2}\sigma_b^2 V'')/N]V(\mu),$$

which, at least for quadratic variance functions, has the same form as the within-blocks variance. Conditionally on the value of the grand mean, $\bar{Y}$, the between-blocks mean square has expectation

$$[\sigma_b^2+\sigma^2(1+\tfrac{1}{2}\sigma_b^2 V'')/N]b[\bar{y};\sigma_b^2+\sigma^2(1+\tfrac{1}{2}\sigma_b^2 V'')/N,k].$$

Thus we may construct an estimator s_b^2 of σ_b^2 such that

$$\{s_b^2+s^2[1+\tfrac{1}{2}s_b^2 V''(\bar{y})]/N\}\,b[\bar{y};s_b^2+s^2(1+\tfrac{1}{2}s_b^2 V''(\bar{y}))/N,k]$$

is equal to the between-blocks mean square. Just as in components-of-variance problems, s_b^2 may turn out to be negative.

The method just described can be extended to more complex designs involving treatment effects where it is important to attribute to each estimated contrast the proper estimate of variance.

If detailed distributional assumptions were made it would generally be possible to produce estimators more efficient than these given here.

Haberman (1982) discusses an analysis of dispersion for multinomial data.

APPENDIX A

The asymptotic distribution of the Poisson deviance for large μ

Let the cumulants κ_r *of* Y be $O(\mu)$ for all r as $\mu \to \infty$. Then $Z = (Y-\mu)/\mu = O_p(\mu^{-\frac{1}{2}})$. Expanding in terms of Z, we have

$$\begin{aligned}&2Y\log(Y/\mu)-2(Y-\mu)\\&\quad = \mu(Z^2 - Z^3/3 + Z^4/6 - Z^5/10 + Z^6/15) + O_p(\mu^{-\frac{5}{2}}).\end{aligned}$$

The mean and variance are given by

$$\begin{aligned}&E\{2Y\log(Y/\mu)-2(Y-\mu)\}\\&\quad = \kappa_2/\kappa_1 + (3\kappa_2^2 - 2\kappa_1\kappa_3)/6\kappa_1^3 + (6\kappa_2^3 - 6\kappa_1\kappa_2\kappa_3 + \kappa_1^2\kappa_4)/6\kappa_1^5 + O(\mu^{-3})\end{aligned}$$

and

$$\begin{aligned}&\text{var}\{2Y\log(Y/\mu)-2(Y-\mu)\}\\&\quad = 2\kappa_2^2/\kappa_1^2 + (3\kappa_1^2\kappa_4 - 18\kappa_1\kappa_2\kappa_3 + 17\kappa_2^3)/3\kappa_1^4 + O(\mu^{-2}),\end{aligned}$$

which, to the orders shown, involve cumulants up to the fourth order. Under the fourth-order assumptions that $\kappa_{r+1} = (\sigma^2)^r\kappa_1$, $r = 1, 2, 3$, these become

$$E\{2Y\log(Y/\mu)-2(Y-\mu)\} = \sigma^2 + \sigma^4/6\mu + \sigma^6/6\mu^2 + O(\mu^{-3})$$

and

$$\begin{aligned}\text{var}\{2Y\log(Y/\mu)-2(Y-\mu)\} &= 2\sigma^4 + 2\sigma^6/3\mu + O(\mu^{-2})\\&= 2[E(2Y\log(Y/\mu)-2(Y-\mu))]^2 + O(\mu^{-2}).\end{aligned}$$

These results suggest that, for large μ and under assumptions of the appropriate order, $[2Y\log(Y/\mu)-2(Y-\mu)]/\sigma^2$ or $[2Y\log(Y/\mu)-2(Y-\mu)]/(\sigma^2+\sigma^4/6\mu)$ has asymptotically the χ_1^2 distribution. Further assumptions, however, are required to establish this result. For example, if cumulants of all orders exist satisfying $\kappa_{r+1} = (\sigma^2)^r\kappa_1$, $r = 1, 2, \ldots$, the result holds, but the result does not appear to hold under corresponding assumptions of finite order.

For Poisson-distributed observations $\sigma^2 = 1$ and the mean and variance of the deviance component are

$$1+\frac{1}{6\mu}+\frac{1}{6\mu^2}+O(\mu^{-3})$$

and
$$2+\frac{2}{3\mu}+O(\mu^{-2})$$

respectively for large μ. It follows from Lawley's (1956) results that, for Poisson-distributed observations and for large μ, the moments of the corrected statistic

$$[2Y\log(Y/\mu)-2(Y-\mu)]/(1+1/6\mu)$$

agree with those of χ_1^2 apart from terms that are $O(\mu^{-2})$.

APPENDIX B

The asymptotic distribution of the binomial deviance

We use the same technique as in Appendix A. Writing $Z = Y - \mu$ and expanding in powers of Z we find

$$2Y\log(Y/\mu) + 2(n-Y)\log[(n-Y)/(n-\mu)]$$
$$= a_1 Z^2 + a_2 Z^3 + a_3 Z^4 + a_4 Z^5 + a_5 Z^6 + O_p(\mu^{-\frac{5}{2}})$$

provided that $Z = O_p(\mu^{\frac{1}{2}})$. This condition is satisfied if the cumulants of Y are all $O(n)$ and $\mu/n \to \Pi$ with $0 < \Pi < 1$ as $n \to \infty$. The coefficients are

$$a_1 = \kappa_1^{-1} + (n-\kappa_1)^{-1}, \qquad 3a_2 = (n-\kappa_1)^{-2} - \kappa_1^{-2},$$
$$6a_3 = \kappa_1^{-3} + (n-\kappa_1)^{-3}, \qquad 10a_4 = (n-\kappa_1)^{-4} - \kappa_1^{-4},$$
$$15a_5 = \kappa_1^{-5} + (n-\kappa_1)^{-5}.$$

The first two moments are

$$a_1\kappa_2 + (a_2\kappa_3 + 3a_3\kappa_2^2) + (a_4\kappa_4 + 10a_5\kappa_2\kappa_3 + 15a_6\kappa_2^3) + O(n^{-3}) \tag{A.1}$$

and

$$2a_1^2\kappa_2^2 + (a_1^2\kappa_4 + 18a_1a_2\kappa_2\kappa_3 + 15a_2^2\kappa_2^3 + 24a_1a_3\kappa_2^3) + O(n^{-2}) \tag{A.2}$$

respectively. The leading term in each case is $O(1)$, the second term being $O(n^{-1})$. Note that, to first order in n and assuming only that the cumulants of Y are $O(n)$, the variance of the deviance component is twice the square of the mean, strongly suggesting a χ_1^2 distribution.

We now make the simplifying assumption that

$$\kappa_2 = \sigma^2\kappa_1(1-\Pi), \qquad \kappa_3 = \sigma^4\kappa_1(1-\Pi)(1-2\Pi),$$
$$\kappa_4 = \sigma^6\kappa_1(1-\Pi)[1-6\Pi(1-\Pi)], \tag{A.3}$$

where $\Pi = \kappa_1/n$ and σ^2 is a measure of over- or under-dispersion relative to the binomial distribution. Under (A.3) the mean and variance to order n^{-1} are

$$\sigma^2 + \sigma^4[1 - \Pi(1-\Pi)]/[6n\Pi(1-\Pi)] \tag{A.4}$$

and

$$2\sigma^4 + 2\sigma^6[1 - \Pi(1-\Pi)]/[3n\Pi(1-\Pi)] \tag{A.5}$$

respectively. Note that under the fourth-order assumptions (A.3) the variance is equal to twice the square of the mean to order n^{-1}. For binomially distributed observations the asymptotic mean and variance are obtained by putting $\sigma^2 = 1$ in (A.4) and (A.5). Lawley's (1956) results show that, for binomially distributed data, the deviance component divided by its mean, $1 + [1 - \Pi(1-\Pi)]/[6n\Pi(1-\Pi)]$, matches closely the moments of χ_1^2 random variable.

Note that the $O(\mu^{-1})$ correction appropriate for Poisson observations with a Poisson log-likelihood function is one-sixth the sum of the reciprocals of the fitted values. This is larger than the $O(\mu^{-1})$ correction appropriate for binomial data. Thus, although the Poisson log-likelihood function, under certain conditions, gives results identical to a binomial log-likelihood function, the goodness-of-fit statistics will differ when terms of order μ^{-1} are taken into account.

APPENDIX C

The asymptotic distribution of the deviance: second-order assumptions

Let $l(\boldsymbol{\beta}; \mathbf{y})$ be the quasi-likelihood function given by

$$l(\boldsymbol{\beta}; \mathbf{y}) = \sum_i l_i(\boldsymbol{\beta}; y_i) = \sum_i [y_i \theta_i - b(\theta_i)], \qquad \text{(A.6)}$$

where $\theta_i = \theta(\mu_i)$ and $b(\boldsymbol{\theta})$ are known functions satisfying the second-order properties

$$\mathrm{E}(Y_i) = \mu_i = b'(\theta_i), \qquad \text{(A.7)}$$

$$\mathrm{cov}(\mathbf{Y}) = \sigma^2 \,\mathrm{diag}[b''(\theta_i)], \qquad \text{(A.8)}$$

where σ^2 may be known or unknown. The vector $\boldsymbol{\mu}$ of length N is expressed as a function of the p-dimensional parameter $\boldsymbol{\beta}$. For generalized linear models this relationship takes the form

$$g(\mu_i) = \sum_j x_{ij} \beta_j; \qquad i = 1, \ldots, N,$$

where the xs are known constants and $g(\ )$ is a known link function. The present results are quite general and apply also to non-linear models provided only that $\boldsymbol{\mu}$ uniquely determines $\boldsymbol{\beta}$. For generalized linear models this condition implies that $\mathbf{X} = \{x_{ij}\}$ has full rank $p \leqslant N$ and that $g(\ )$ is invertible.

The function $l(\boldsymbol{\beta}; \mathbf{y})$ together with its matrix of second derivatives

$$-\mathbf{I}_\beta(\boldsymbol{\beta}, \mathbf{y}) = \frac{\partial^2 l}{\partial \boldsymbol{\beta}^2} \qquad \text{and} \qquad \mathbf{i}_\beta = \mathrm{E}(\mathbf{I}_\beta(\boldsymbol{\beta}, \mathbf{Y}))$$

are assumed to be of order N: the first derivative vector, $\mathbf{u}_\beta(\boldsymbol{\beta}, \mathbf{y})$, is of order $N^{\frac{1}{2}}$ with zero expectation. Note that, since $l(\boldsymbol{\beta}; \mathbf{y})$ is linear in $\mathbf{y}$, only first-order assumptions are required to compute $\mathrm{E}(\mathbf{I}_\beta(\boldsymbol{\beta}; \mathbf{Y}))$.

The following argument, which assumes a degree of regularity, demonstrates that for large n there exists a consistent solution, $\hat{\boldsymbol{\beta}}$, of the quasi-likelihood equations. Note that $\hat{\boldsymbol{\beta}}$ is not the maximum-likelihood estimate of $\boldsymbol{\beta}$ but we will see that $\hat{\boldsymbol{\beta}}-\boldsymbol{\beta} = O_p(N^{-\frac{1}{2}})$, implying consistency. Furthermore, $\mathbf{I}_\beta - \mathbf{i}_\beta = O_p(N^{\frac{1}{2}})$, in general, although for certain generalized linear models the difference is identically zero. Strictly speaking we should distinguish $\mathbf{i}_\beta(\boldsymbol{\beta})$ and $\mathbf{i}_\beta(\hat{\boldsymbol{\beta}})$. However, we suppress the argument $\boldsymbol{\beta}$ or $\hat{\boldsymbol{\beta}}$ because, under appropriate assumptions of continuity, these quantities differ by a term of sufficiently small order.

The equations used for determining $\hat{\boldsymbol{\beta}}$, $\mathbf{u}_\beta(\hat{\boldsymbol{\beta}}; \mathbf{y}) = 0$, can be written

$$\begin{aligned} \mathbf{0} &= \mathbf{u}_\beta(\boldsymbol{\beta}; \mathbf{y}) - \mathbf{I}_\beta(\boldsymbol{\beta}; \mathbf{y})(\hat{\boldsymbol{\beta}}-\boldsymbol{\beta}) + O_p(1) \\ &= \mathbf{u}_\beta(\boldsymbol{\beta}; \mathbf{y}) - \mathbf{i}_\beta(\hat{\boldsymbol{\beta}}-\boldsymbol{\beta}) + O_p(1) \qquad (A.9) \\ \hat{\boldsymbol{\beta}}-\boldsymbol{\beta} &= \mathbf{i}_\beta^{-1}\mathbf{u}_\beta(\boldsymbol{\beta}; \mathbf{y}) + O_p(N^{-1}). \end{aligned}$$

Equation (A.9) describes the iterative scheme for computing $\hat{\boldsymbol{\beta}}$. The term on the right is of order $N^{-\frac{1}{2}}$, showing that $\hat{\boldsymbol{\beta}}$ is consistent for $\boldsymbol{\beta}$. Note that $\mathbf{i}(\boldsymbol{\beta})$ is not the Fisher information matrix because only second-order assumptions have been made concerning the distribution of $\mathbf{Y}$.

Using a Taylor series expansion for $l(\boldsymbol{\beta}; \mathbf{y})$ about $\hat{\boldsymbol{\beta}}$ we find that

$$l(\boldsymbol{\beta}; \mathbf{y}) - l(\hat{\boldsymbol{\beta}}; \mathbf{y}) = -\tfrac{1}{2}(\hat{\boldsymbol{\beta}}-\boldsymbol{\beta})\mathbf{i}_\beta(\hat{\boldsymbol{\beta}}-\boldsymbol{\beta}) + O_p(N^{-\frac{1}{2}}). \qquad (A.10)$$

Substituting (A.9) and emphasizing that the quantities $\hat{\boldsymbol{\beta}}$ and Y are considered random, we have that

$$2l(\hat{\boldsymbol{\beta}}; \mathbf{Y}) - 2l(\boldsymbol{\beta}; \mathbf{Y}) = \mathbf{u}_\beta(\boldsymbol{\beta}; \mathbf{Y})\mathbf{i}_\beta^{-1}\mathbf{u}_\beta(\boldsymbol{\beta}; \mathbf{Y}) + O_p(N^{-\frac{1}{2}}). \qquad (A.11)$$

Now $\mathbf{u}_\beta(\boldsymbol{\beta}; \mathbf{Y})$ involves a sum over the data vector so that a strong central-limit effect applies. Furthermore, by (A.8) $\text{cov}(\mathbf{u}_\beta(\boldsymbol{\beta}; \mathbf{Y})) = \sigma^2\mathbf{i}(\boldsymbol{\beta})$ exactly. It follows therefore that

$$2l(\hat{\boldsymbol{\beta}}; \mathbf{Y}) - 2l(\boldsymbol{\beta}; \mathbf{Y}) \sim \sigma^2\chi^2_p + O_p(N^{-\frac{1}{2}}). \qquad (A.12)$$

Result (A.12) is a property of the function $l(\boldsymbol{\beta}; \mathbf{Y})$: the limiting χ^2 distribution applies regardless of the distribution of the data provided only that the second-moment properties (A.7) and (A.8) hold for $\mathbf{Y}$.

Written in terms of the deviance function, $D(\mathbf{y}; \boldsymbol{\mu})$, the limiting result (A.12) becomes

$$D(\mathbf{Y}; \boldsymbol{\mu}) - D(\mathbf{Y}; \hat{\boldsymbol{\mu}}) \sim \sigma^2\chi^2_p + O_p(N^{-\frac{1}{2}}). \qquad (A.13)$$

We cannot, unfortunately, infer from (A.13) that $D(\mathbf{Y}; \hat{\boldsymbol{\mu}}) \sim \chi^2_{N-p}$.

However, the mean of $D(\mathbf{Y};\hat{\boldsymbol{\mu}})$ is given by

$$\mathrm{E}(D(\mathbf{Y};\hat{\boldsymbol{\mu}})) = \mathrm{E}(D(\mathbf{Y};\boldsymbol{\mu}))-p\sigma^2+\mathrm{O}(N^{-1}). \qquad \text{(A.14)}$$

Usually the second-moment assumptions (A.7) and (A.8) are not sufficient to calculate $\mathrm{E}(D(\mathbf{Y};\boldsymbol{\mu}))$: detailed knowledge of further aspects of the distribution of $\mathbf{Y}$ is usually required. The variance of $D(\mathbf{Y};\hat{\boldsymbol{\mu}})$ involves cumulants of $\mathbf{Y}$ at least up to the fourth order.

Nuisance parameters

Let the parameter vector $\boldsymbol{\beta} = (\boldsymbol{\psi}, \boldsymbol{\lambda})$ of total length p be partitioned into two components, where $\boldsymbol{\psi}$ is the parameter vector of interest and $\boldsymbol{\lambda}$ is a nuisance-parameter vector of length q. Suppose we wish to test the composite null hypothesis $\mathrm{H}_0: \boldsymbol{\psi} = \boldsymbol{\psi}_0$, the remaining q parameters being unspecified. The previous results apply both under H_0 and under $\mathrm{H}_0: \boldsymbol{\psi} \neq \boldsymbol{\psi}_0$. Let the two sets of estimates be $(\boldsymbol{\psi}_0, \hat{\boldsymbol{\lambda}}_0)$ under H_0 and $(\hat{\boldsymbol{\psi}}, \hat{\boldsymbol{\lambda}})$ under H_A. The asymptotic covariance matrix, $\sigma^2\mathbf{i}_{\beta}^{-1}$, of $\hat{\boldsymbol{\beta}}$ is partitioned according to the partition $(\boldsymbol{\psi}, \boldsymbol{\lambda})$ of $\boldsymbol{\beta}$ and are given by

$$\sigma^2\begin{bmatrix}\mathbf{0} & \mathbf{0}\\ \mathbf{0} & \mathbf{i}_{\lambda\lambda}^{-1}\end{bmatrix} \quad \text{and} \quad \sigma^2\begin{bmatrix}\mathbf{i}_{\psi\psi} & \mathbf{i}_{\psi\lambda}\\ \mathbf{i}_{\lambda\psi} & \mathbf{i}_{\lambda\lambda}\end{bmatrix}^{-1}$$

respectively. By the standard property of the limiting Normal density the restricted maximum $\hat{\psi}_0$ is given by

$$\hat{\boldsymbol{\lambda}}_0 = \hat{\boldsymbol{\lambda}}+\mathbf{i}_{\lambda\lambda}^{-1}\mathbf{i}_{\lambda\psi}(\hat{\boldsymbol{\psi}}-\boldsymbol{\psi}_0), \qquad \text{(A.15)}$$

where $-\mathbf{i}_{\lambda\lambda}^{-1}\mathbf{i}_{\lambda\psi}$ is the matrix of regression coefficients of $\hat{\boldsymbol{\lambda}}$ on $\boldsymbol{\psi}$. Using (A.10) and neglecting terms of order $N^{-\frac{1}{2}}$, the test statistic $2l(\hat{\boldsymbol{\psi}}, \hat{\boldsymbol{\lambda}}; \mathbf{y})-2l(\boldsymbol{\psi}_0, \hat{\boldsymbol{\lambda}}_0)$ becomes

$$\begin{bmatrix}\hat{\boldsymbol{\psi}}-\boldsymbol{\psi}_0\\ \hat{\boldsymbol{\lambda}}-\boldsymbol{\lambda}\end{bmatrix}^{\mathrm{T}}\begin{bmatrix}\mathbf{i}_{\psi\psi} & \mathbf{i}_{\psi\lambda}\\ \mathbf{i}_{\lambda\psi} & \mathbf{i}_{\lambda\lambda}\end{bmatrix}\begin{bmatrix}\hat{\boldsymbol{\psi}}-\boldsymbol{\psi}_0\\ \hat{\boldsymbol{\lambda}}-\boldsymbol{\lambda}\end{bmatrix}-(\hat{\boldsymbol{\lambda}}_0-\boldsymbol{\lambda})^{\mathrm{T}}\mathbf{i}_{\lambda\lambda}(\hat{\boldsymbol{\lambda}}_0-\boldsymbol{\lambda}).$$

Substituting (A.15) this reduces to

$$(\hat{\boldsymbol{\psi}}-\boldsymbol{\psi}_0)\{\mathbf{i}_{\psi\psi}-\mathbf{i}_{\psi\lambda}\mathbf{i}_{\lambda\lambda}^{-1}\mathbf{i}_{\lambda\psi}\}(\hat{\boldsymbol{\psi}}-\boldsymbol{\psi}_0). \qquad \text{(A.16)}$$

The matrix of the quadratic form (A.16) is $(\mathbf{i}^{\psi\psi})^{-1}$ which is proportional to the inverse of the limiting covariance matrix of $\hat{\boldsymbol{\psi}}$. Since the limiting distribution of $\hat{\boldsymbol{\psi}}$ is normal, it follows that

$$2l(\hat{\boldsymbol{\psi}}, \hat{\boldsymbol{\lambda}}; \mathbf{Y})-2l(\boldsymbol{\psi}_0, \hat{\boldsymbol{\lambda}}_0; \mathbf{Y}) \sim \sigma^2\chi^2_{p-q}+\mathrm{O}_p(N^{-\frac{1}{2}}).$$

Written in terms of the deviances $D(\mathbf{y};\hat{\boldsymbol{\mu}})$ and $D(\mathbf{Y};\hat{\boldsymbol{\mu}}_0)$ we have that

$$D(\mathbf{Y};\hat{\boldsymbol{\mu}}_0)-D(\mathbf{Y};\hat{\boldsymbol{\mu}}) \sim \sigma^2\chi^2_{p-q}+O_p(N^{-\frac{1}{2}}). \qquad \text{(A.17)}$$

These results assume that H_0 is in fact true.

The second-moment assumptions (A.7) and (A.8) can be generalized to permit correlated observations. Let $\boldsymbol{\mu}$ and $\sigma^2\mathbf{V}$ be the mean vector and covariance matrix of $\mathbf{Y}$. If we can find a function $b(\boldsymbol{\theta})$ with $b'(\boldsymbol{\theta}) = \boldsymbol{\mu}$ and $b''(\boldsymbol{\theta}) = \mathbf{V}$, where $\boldsymbol{\theta}$ is of length N, the quasi-likelihood function is $\mathbf{y}^T\boldsymbol{\theta}-b(\boldsymbol{\theta})$. The results (A.13) and (A.17) still apply provided that a version of the central-limit theorem applies to $\mathbf{u}_\beta(\boldsymbol{\beta};\mathbf{Y})$.

APPENDIX D

The asymptotic distribution of the deviance: a higher-order approximation

Lawley (1956) has shown how to modify the likelihood-ratio statistic so that the modified statistic behaves like χ^2 when terms of order N^{-2} are ignored. His results are not directly applicable here because we work with quasi-likelihood functions and not with genuine log-likelihood functions. Nevertheless, with slight modifications, essentially the same calculations can be used here with inferences based on the quasi-likelihood function

$$l(\boldsymbol{\beta}; \mathbf{y}) = \sum_i [y_i \theta_i - b(\theta_i)]. \tag{A.18}$$

As before, $\theta_i = \theta(\mu_i)$ and $b(\theta)$ are known functions chosen to mimic the second-moment properties of $\mathbf{Y}$:

$$\boldsymbol{\mu} = \mathrm{E}(\mathbf{Y}) = \{b'(\theta_1), \ldots, b'(\theta_N)\}^{\mathrm{T}}, \qquad \mathrm{cov}(\mathbf{Y}) = \sigma^2 \,\mathrm{diag}[b''(\theta_i)]. \tag{A.19}$$

In order to carry the approximations to a higher order, we make additional fourth-moment assumptions concerning the cumulants $\kappa_3^{(i)}$ and $\kappa_4^{(i)}$ of Y_i. These are

$$\kappa_3^{(i)} = \sigma^4 b'''(\theta_i) \qquad \text{and} \qquad \kappa_4^{(i)} = \sigma^6 b^{(\mathrm{iv})}(\theta_i). \tag{A.20}$$

In addition we require fourth-order independence defined by

$$\mathrm{E}((Y_i-\mu_i)(Y_j-\mu_j)(Y_k-\mu_k)) = \begin{cases} \kappa_3^{(i)}; & i=j=k, \\ 0; & \text{otherwise,} \end{cases} \tag{A.21a}$$

and

$$\mathrm{E}((Y_i-\mu_i)(Y_j-\mu_j)(Y_k-\mu_k)(Y_l-\mu_l)) = \begin{cases} \kappa_4^{(i)} + 3(\kappa_2^{(i)})^2; & i=j=k=l, \\ \kappa_2^{(i)}\kappa_2^{(k)}; & i=j \neq k=l, \\ 0; & \text{otherwise.} \end{cases} \tag{A.21b}$$

Properties (A.20) and (A.21) would automatically be satisfied if $l(\boldsymbol{\beta}; \mathbf{y})$ were proportional to the log-likelihood function for $\mathbf{Y}$. The essential point here is that if a quasi-likelihood function, whose fourth-moment properties mimic those of the data, can be found, a higher order of approximation to the distribution of the likelihood-ratio statistic is possible, at least in principle.

As usual the parameter vector $\boldsymbol{\beta}$ of dimension p specifies the relation between the mean vector $\boldsymbol{\mu}$ and the set of covariates or experimental conditions $\mathbf{X}$. For the purpose of the present discussion this relation need not be linear but $\boldsymbol{\mu}$ must uniquely determine $\boldsymbol{\beta}$. Let $\boldsymbol{\beta} = (\boldsymbol{\psi}, \boldsymbol{\lambda})$ be partitioned into two components of dimensions $p-q$ and q respectively. The composite null hypothesis H_0 specifies $\boldsymbol{\psi} = \boldsymbol{\psi}_0$: the q parameters λ are nuisance parameters. Maximum-quasi-likelihood estimates are denoted by $(\boldsymbol{\psi}_0, \hat{\boldsymbol{\lambda}}_0)$ under H_0 and by $(\hat{\boldsymbol{\psi}}, \hat{\boldsymbol{\lambda}})$ under H_A. The corresponding fitted values are $\hat{\boldsymbol{\mu}}_0$ and $\hat{\boldsymbol{\mu}}$ respectively. To test H_0 we use the statistic $D(\mathbf{Y}; \hat{\boldsymbol{\mu}}_0) - D(\mathbf{Y}; \hat{\boldsymbol{\mu}})$. Under H_0 and neglecting terms of order N^{-2}, this statistic has expectation

$$E = \sigma^2(p-q) + \epsilon\sigma^4/N$$

where ϵ depends on the first three cumulants and on the form of the systematic part of the model. Explicit formulae for ϵ appropriate to generalized linear models have been given by Cordeiro (1983a, b). The general form for ϵ, even for generalized linear models, is rather cumbersome: in applications it is often easier to compute expectations directly rather than resort to substituting into general formulae.

The variation of Lawley's result appropriate to deviances based on quasi-likelihood functions obeying the fourth-moment assumptions (A.20) and (A.21) is that

$$(p-q)[D(\mathbf{Y}; \hat{\boldsymbol{\mu}}_0) - D(\mathbf{Y}; \hat{\boldsymbol{\mu}})]/E \sim \chi^2_{p-q} + O_p(N^{-2}). \quad \text{(A.22)}$$

The factor $E = 1 + \epsilon\sigma^2/N(p-q)$ is known as Bartlett's correction factor after Bartlett (1937, 1954). The computation of ϵ involves only cumulants up to the third but the validity of (A.22) requires also fourth-moment assumptions. For large N the correction factor reduces to unity giving result (A.17). For moderate values of N result (A.22) depends heavily on the second-moment assumptions (A.19) but only weakly on the third- and fourth-moment assumptions (A.20) and (A.21). Consequently, in applications, the exactness of (A.20) and (A.21) appears not to be crucial.

APPENDIX E

Software

Two computer packages have been designed explicitly to handle the fitting of generalized linear models by maximum likelihood; these are Genstat and GLIM. In Genstat the original regression section was expanded to deal with generalized linear models by including additional options for error distributions and link functions. Originally a batch program, Genstat can now be set up to work interactively, and it contains a useful summary printing facility whereby the results of fitting a sequence of models can be displayed in tabular form. GLIM was designed to work interactively from the beginning and is intended for exploratory fitting; it includes ancillary operations like sorting, plotting and editing but is a smaller program, having fewer internal data structures and simpler forms of output, than Genstat. Both packages have interpretive languages of considerable generality and include control structures such as macros, branching and looping. They both use the model formulae of Chapter 3 for the specification of linear predictors.

Further information on the above packages can be obtained from the distributors:

Numerical Algorithms Group Limited
NAG Central Office
Mayfield House
256 Banbury Road
Oxford
England

Various programs have been produced which allow the fitting of particular cases of generalized linear models; mostly they cover classical regression or log–linear models for counts. Some of the former have been reviewed by Searle (1979); it is usually either difficult or impossible to adapt them easily to deal with generalized linear models. Log–linear models constitute a special case because of the existence of the iterative proportional scaling algorithm for fitting them. ECTA is an example of a program using this algorithm. The program GENCAT uses weighted regression with empirical weights, yielding results very similar to a single cycle of the maximum-likelihood algorithm of Genstat and GLIM. Special programs have been developed for survival data involving censored observations and time-dependent covariates by Byar *et al.* and others. For summaries of statistical software see Francis (1981).

References

Adena, M.A. and Wilson, S.R. (1982) *Generalized Linear Models in Epidemiological Research, INTSTAT, Sydney.*

Agresti, A. (1977) Considerations in measuring partial association for ordinal categorical data. *J. Am. Statist. Assoc.*, **72**, 32–45.

Aitkin, M. and Clayton, D. (1980) The fitting of exponential, Weibull and extreme value distributions to complex censored survival data using GLIM. *Appl. Statist.*, **29**, 156–63.

Akaike, H. (1969) Fitting autoregressive models for prediction. *Ann. Inst. Statist. Math.*, **21**, 243–7.

Akaike, H. (1973) Information theory and an extension of the maximum likelihood principle, in *Second International Symposium on Information Theory* (ed. B.N. Petrov and F. Czáki), Akademiai Kiadó, Budapest, pp. 267–81.

Andersen, E.B. (1973) *Conditional Inference and Models for Measuring*, Mentalhygiejnisk Forlag, Copenhagen.

Andrews, D.F. and Pregibon, D. (1978) Finding the outliers that matter. *J. R. Statist. Soc.*, B, **40**, 85–93.

Anscombe, F.J. (1949) The statistical analysis of insect counts based on the negative binomial distribution. *Biometrics*, **15**, 165–73.

Anscombe, F.J. (1953) Contribution to the discussion of H. Hotelling's paper. *J. R. Statist. Soc.*, B, **15**, 229–30.

Anturane Reinfarction Trial Research Group (1978) Sulfinpyrazone in the prevention of cardiac death after myocardial infarction. *New Engl. J. Med.*, **298**, 289–95.

Anturane Reinfarction Trial Research Group (1980) Sulfinpyrazone in the prevention of sudden death after myocardial infarction. *New Engl. J. Med.*, **302**, 250–6.

Armitage, P. (1957) Studies in the variability of pock counts. *J. Hygiene Camb.*, **55**, 564–81.

Armitage, P. (1971) *Statistical Methods in Medical Research*, Blackwell, Oxford.

Ashford, J.R. (1959) An approach to the analysis of data for semi-quantal responses in biological assay. *Biometrics*, **15**, 573–81.

Ashton, W.D. (1972) *The Logit Transformation with Special Reference to its Uses in Bioassay*, Griffin, London.

Atkinson, A.C. (1981 a) Two graphical displays for outlying and influential observations in regression. *Biometrika*, **68**, 13–20.

Atkinson, A.C. (1981b) Likelihood ratios, posterior odds and information criteria. *J. Econometrics*, **16**, 15–20.

Atkinson, A.C. (1982). Regression diagnostics, transformations and constructed variables (with discussion). *J. R. Statist. Soc.*, B, **44**, 1–36.

Baker, R.J. and Nelder, J. A. (1978) *The GLIM System*, Release 3, *Generalized Linear Interactive Modelling*, Numerical Algorithms Group, Oxford.

Barndorff-Nielsen, O. (1978) *Information and Exponential Families in Statistical Theory*, J. Wiley & Sons, New York.

Barndorff-Nielsen, O. (1980) Conditionality resolutions. *Biometrika*, **67**, 293–310.

Barndorff-Nielsen, O. and Cox, D.R. (1979) Edgeworth and saddle-point approximations with statistical applications. *J. R. Statist. Soc.*, B, **41**, 279–312.

Bartlett, M.S. (1936) Some notes on insecticide tests in the laboratory and in the field. *J. R. Statist. Soc.*, **3**, Suppl. 185–94.

Bartlett, M.S. (1937) Properties of sufficiency and statistical tests. *Proc. R. Soc.*, A, **160**, 268–82.

Bartlett, M.S. (1954) A note on the multiplying factors for various χ^2 approximations. *J. R. Statist. Soc.*, B, **16**, 296–8.

Baxter, L.A., Coutts, S.M. and Ross, G.A. F. (1980) Applications of linear models in motor insurance, in *Proceedings of the 21st International Congress of Actuaries*, Zurich, pp. 11–29.

Beale, E.M.L. (1970) Note on procedures for variable selection in multiple regression. *Technometrics*, **12**, 909–14.

Beaton, A.E. (1964) The use of special matrix operators in statistical calculus. *Research Bulletin RB-64-51, Educational Testing Service*, Princeton, New Jersey, USA.

Belsley, D.A., Kuh, E. and Welsch, R.E. (1980) *Regression Diagnostics*, J. Wiley & Sons, New York.

Berge, C. (1973) *Graphs and Hypergraphs*, North-Holland, Amsterdam. Translated and revised from *Graphs et Hypergraphs*, Dunod, Paris (1970).

Berk, R.H. (1972). Consistency and asymptotic normality of MLE's for exponential models. *Ann. Math. Statist.*, **43**, 193–204.

Berkson, J. (1944) Application of the logistic function to bio-assay. *J. Am. Statist. Assoc.*, **39**, 357–65.

Berkson, J. (1951) Why I prefer logits to probits. *Biometrics*, **7**, 327–39.

Bhapkar, V.P. (1966) A note on the equivalence of two test criteria for hypotheses in categorical data. *J. Am. Statist. Assoc.*, **61**, 228–35.

Birch, M.W. (1965) The detection of partial association II: the general case. *J. R. Statist. Soc.*, B, **27**, 111–24.

Bishop, Y.M.M., Fienberg, S.E. and Holland, P.W. (1975) *Discrete Multivariate Analysis: Theory and Practice*, MIT Press, Cambridge, MA.

Björck, Å. (1967) Solving linear least squares problems by Gram–Schmidt orthogonalization. *BIT*, **7**, 257–78.

Bliss, C.I. (1935) The calculation of the dosage–mortality curve. *Ann. Appl. Biol.*, **22**, 134–67.

Bliss, C.I. (1970) *Statistics in Biology*, vol. II, McGraw-Hill, New York.

Bock, R.D. (1975) *Multivariate Statistical Methods in Behavioral Research*, McGraw-Hill, New York.

Bortkewitsch, L. (1898) *Das Gesetz der Kleiner Zaahlen*, Teubner, Leipzig.

Box, G.E.P. (1980) Sampling and Bayes' inference in scientific modelling and robustness. *J. R. Statist. Soc.*, A, **143**, 383–430.

Box, G.E.P. and Cox, D.R. (1964) An analysis of transformations. *J. R. Statist. Soc.*, B, **26**, 211–52.

Box, G.E.P. and Jenkins, G.M. (1976) *Time Series Analysis: Forecasting and Control*, Holden-Day, San Francisco.

Box, G.E.P. and Tidwell, P.W. (1962) Transformation of the independent variables. *Technometrics*, **4**, 531–50.

Breslow, N. (1976) Regression analysis of the log odds ratio: a method for retrospective studies. *Biometrics*, **32**, 409–16.

Breslow, N. (1981) Odds ratio estimators when the data are sparse. *Biometrika*, **68**, 73–84.

Breslow, N. and Day, N.E. (1980) *Statistical Methods in Cancer Research*, vol. 1, *The Analysis of Case-Control Studies*, IARC, Lyon.

Bross, I.D.J. (1958) How to use ridit analysis. *Biometrics*, **14**, 18–38.

Byar, D.T. (1983) Analysis of survival data: Cox and Weibull models with covariates, in *Statistics in Medical Research: Methods and Issues, with Applications in Clinical Oncology* (ed. V. Mike and K. Stanley), J. Wiley & Sons, New York.

Caussinus, H. (1965) Contribution à l'analyse statistique des tableaux de correlation. *Ann. Fac. Sci. Univ. Toulouse*, **29**, 77–182.

Chambers, E.A. and Cox, D.R. (1967) Discrimination between alternative binary response models. *Biometrika*, **54**, 573–8.

Chambers, J.M. (1977) *Computational Methods for Data Analysis*, J. Wiley & Sons, New York.

Clarke, M.R.B. (1981) Algorithm AS163: a Givens algorithm for moving from one linear model to another without going back to the data. *Appl. Statist.*, **30**, 198–203.

Clarke, M.R.B. (1982) Algorithm AS178: the Gauss–Jordan sweep operator with detection of collinearity. *Appl. Statist.*, **31**, 166–8.

Clayton, D.G. (1974) Some odds ratio statistics for the analysis of ordered categorical data. *Biometrika*, **61**, 525–31.

Coe, R. and Stern, R.D. (1982) Fitting models to daily rainfall. *J. Appl. Meteorol.*, **21**, 1024–31.

Cook, R.D. (1977) Detection of influential observations in linear regression. *Technometrics*, **19**, 15–18.

Cook, R.D. (1979) Influential observations in linear regression. *J. Am. Statist. Assoc.*, **74**, 169–74.

Cook, R.D. and Weisberg, S. (1982) *Residuals and Influence in Regression*, Chapman and Hall, London.

Cordeiro, G.M. (1983a) Improved likelihood-ratio tests for generalized linear models. *J. R. Statist. Soc.*, B, **45**, 404–13.

Cordeiro, G.M. (1983b) Improved likelihood-ratio tests for generalized linear models. *PhD Thesis*, University of London.

Cox, D.R. (1958a) The regression analysis of binary sequences (with discussion). *J. R. Statist. Soc.*, B, **20**, 215–42.

Cox, D.R. (1958b) Two further applications of a model for binary regression. *Biometrika*, **45**, 562–5.

Cox, D.R. (1962) *Renewal Theory*, Chapman and Hall, London.

Cox, D.R. (1970) *The Analysis of Binary Data*, Chapman and Hall, London.

Cox, D.R. (1972) Regression models and life tables (with discussion). *J. R. Statist. Soc.*, B, **74**, 187–220.

Cox, D.R. (1975) Partial likelihood. *Biometrika*, **62**, 269–76.

Cox, D.R. and Hinkley, D.V. (1968) A note on the efficiency of least-squares estimates. *J. R. Statist. Soc.*, B, **30**, 284–9.

Cox, D.R. and Hinkley, D.V. (1974) *Theoretical Statistics*, Chapman and Hall, London.

Cox, D.R. and Snell, E.J. (1968) A general definition of residuals. *J. R. Statist. Soc.*, B, **30**, 248–75.

Cox, D.R. and Snell, E.J. (1981) *Applied Statistics; Principles and Examples*, Chapman and Hall, London.

Crowley, J. and Hu, M. (1977) Covariance analysis of heart transplant data. *J. Am. Statist. Assoc.*, **72**, 27–36.

Daniel, C. (1959) Use of half-normal plots for interpreting factorial two-level experiments. *Technometrics*, **1**, 311–41.

Darby, S.C. and Ellis, M.J. (1976) A test for synergism between two drugs. *Appl. Statist.*, **25**, 296–9.

Darroch, J.N., Lauritzen, S.L. and Speed, T.P. (1980) Markov fields and log–linear interaction models for contingency tables. *Ann. Statist.*, **8**, 522–39.

Darroch, J.N. and Ratcliff, D. (1972) Generalized iterative scaling for log–linear models. *Ann. Math. Statist.*, **43**, 1470–80.

Day, N.E. and Byar, D.P. (1979) Testing hypotheses in case-control studies: equivalence of Mantel–Haenszel statistics and logit score tests. *Biometrics*, **35**, 623–30.

Deming, W.E. and Stephan, F.F. (1940) On a least squares adjustment of a sampled frequency table when the expected marginal totals are known. *Ann. Math. Statist.*, **11**, 427–44.

Dempster, A.P. (1971) An overview of multivariate data analysis. *J. Mult. Analysis*, **1**, 316–46.

Dempster, A.P., Laird, N.M. and Rubin, D.B. (1977) Maximum likelihood from incomplete data via the EM algorithm (with discussion). *J. R. Statist. Soc.*, B, **39**, 1–38.

Draper, N.R. and Smith, H. (1981) *Applied Regression Analysis*, 2nd edn, J. Wiley & Sons, New York.

Durbin, J. (1980) Approximations for densities of sufficient estimators. *Biometrika*, **67**, 311–33.

Dyke, G.V. and Patterson, H.D. (1952) Analysis of factorial arrangements when the data are proportions. *Biometrics*, **8**, 1–12.

Efron, B. (1977). The efficiency of Cox's likelihood function for censored data. *J. Am. Statist. Assoc.*, **72**, 557–65.

Efroymson, M.A. (1960) Multiple regression analysis, in *Mathematical*

Methods for Digital Computers, vol. 1 (ed. A. Ralston and H.S. Wilf), J. Wiley & Sons, New York, pp. 191–203.
Elandt-Johnson, R.C. and Johnson, N.H. (1980) *Survival Models and Data Analysis*, J. Wiley & Sons, New York.
Everitt, B.S. (1977) *The Analysis of Contingency Tables*, Chapman and Hall, London.
Fienberg, S.E. (1970a) An iterative procedure for estimation in contingency tables. *Ann Math. Statist.*, **41**, 907–17.
Fienberg, S.E. (1970b) The analysis of multidimensional contingency tables. *Ecology*, **51**, 419–33.
Fienberg, S.E. (1980) *The Analysis of Cross-Classified Data*, 2nd edn, MIT Press, Cambridge, MA.
Finney, D.J. (1971) *Probit Analysis*, 3rd edn, Cambridge University Press, Cambridge.
Finney, D.J. (1976) Radioligand assay. *Biometrics*, **32**, 721–40.
Firth, D. (1982) Estimation of voter transition matrices. *MSc Thesis*, University of London.
Fisher R.A. (1922) On the mathematical foundations of theoretical statistics. *Phil. Trans. R. Soc.*, **222**, 309–68.
Fisher, R.A. (1935) The case of zero survivors (Appendix to Bliss, C.I. (1935)). *Ann. Appl. Biol.*, **22**, 164–5.
Fisher, R.A. (1949) A biological assay of tuberculins. *Biometrics*, **5**, 300–16.
Fleiss, J.L. (1981) *Statistical Methods for Rates and Proportions*, J. Wiley & Sons, New York.
Francis, I. (1981) *Statistical Software: A Comparative Review*, North Holland, New York.
Freireich, E.J. *et al.* (1963) The effect of 6-mercaptopurine on the duration of steroid-induced remissions in acute leukemia. *Blood*, **21**, 699–716.
Furnival, G.M. and Wilson, R.W. (1974) Regression by leaps and bounds. *Technometrics*, **16**, 499–511.
Gail, M.H., Lubin, J.H. and Rubinstein, L.V. (1981) Likelihood calculations for matched case-control studies and survival studies with tied death times. *Biometrika*, **68**, 703–7.
Gart, J.J. and Zweifel, J.R. (1967) On the bias of various estimators of the logit and its variance with applications to quantal bioassay. *Biometrika*, **54**, 181–7.
Gehan, E.A. (1965) A generalized Wilcoxon test for comparing arbitrarily single-censored samples. *Biometrika*, **52**, 203–23.
Gentleman, W.M. (1974a) Regression problems and the QR decomposition. *Bull. I.M.A.*, **10**, 195–7.
Gentleman, W.M. (1974b) Basic procedures for large, sparse or weighted linear least squares problems (Algorithm AS75). *Appl. Statist.*, **23**, 448–57.
Gokhale, D.V. and Kullback, S. (1978) *The Information in Contingency Tables*, Marcel Dekker, New York.
Goodman, L.A. (1973) The analysis of multidimensional contingency tables when some variables are posterior to others: a modified path analysis approach. *Biometrika*, **60**, 179–92.

Goodman, L.A. (1978) *Analysing Qualitiative/Categorical Data*, Addison Wesley, London.

Goodman, L.A. (1979) Simple models for the analysis of association in cross-classifications having ordered categories. *J. Am. Statist. Assoc.*, **74**, 537–52.

Goodman, L.A. (1981) Association models and canonical correlation in the analysis of cross-classifications having ordered categories. *J. Am. Statist. Assoc.*, **76**, 320–34.

Goodman, L.A. and Kruskal, W.H. (1954, 1959, 1963, 1972) Measures of association for cross classification, Parts I–IV. *J. Am. Statist. Assoc.*, **49**, 732–64 (corr. **52**, 578); **54**, 123–63; **58**, 310–64; **67**, 415–21.

Gross, A.J. and Clark, V.A. (1975) *Survival Distributions: Reliability Applications in the Biomedical Sciences*, J. Wiley & Sons, New York.

Gurland, J., Lee, I. and Dahm, P. A. (1960) Polychotomous quantal response in biological assay. *Biometrics*, **16**, 382–98.

Haberman, S.J. (1974a) *The Analysis of Frequency Data*, University of Chicago Press, Chicago.

Haberman, S.J. (1974b) Log–linear models for frequency tables with ordered classifications. *Biometrics*, **36**, 589–600.

Haberman, S.J. (1977) Maximum likelihood estimates in exponential response models. *Ann. Statist.*, **5**, 815–41.

Haberman, S.J. (1982) The analysis of dispersion of multinomial responses. *J. Am. Statist. Assoc.*, **77**, 568–80.

Harter, H.L. (1976) The method of least squares and some alternatives. *Int. Statist. Rev.*, **44**, 113–59.

Hawkins, D. (1980) *Identification of Outliers*, Chapman and Hall, London.

Healy, M.J.R. (1968) Triangular decomposition of a symmetric matrix (Algorithm AS6); Inversion of a positive semi-definite symmetric matrix (Algorithm AS7). *Appl. Statist.*, **17**, 192–7.

Healy, M.J.R. (1981) A source of assay heterogeneity. *Biometrics*, **37**, 834–5.

Hewlett, P.S. (1969) Measurement of the potencies of drug mixtures. *Biometrics*, **25**, 477–87.

Hewlett, P.S. and Plackett, R.L. (1956) The relation between quantal and graded responses to drugs. *Biometrics*, **12**, 72–8.

Hoaglin, D.C. and Welsch, R.E. (1978) The hat matrix in regression and ANOVA. *Am. Statistician*, **32**, 17–22 (Corr. **32**, 146).

Hougaard, P. (1982) Parametrizations of non-linear models. *J. R. Statist. Soc.*, B, **44**, 244–52.

Hurn, M.W., Barker, N.W. and Magath, T.D. (1945) The determination of prothrombin time following the administration of dicumarol with specific reference to thromboplastin. *J. Lab. Clin. Med.*, **30**, 432–47.

Kalbfleisch, J.D. and Prentice, R.L. (1980) *The Statistical Analysis of Failure Time Data*, J. Wiley & Sons, New York.

Killion, R.A. and Zahn, D.A. (1976) A bibliography of contingency table literature. *Int. Statist. Rev.*, **44**, 71–112.

Lane, P.W. and Nelder, J.A. (1982) Analysis of covariance and standardization as instances of prediction. *Biometrics*, **38**, 613–21.

Lawley, D.N. (1956) A general method for approximating to the distribution of likelihood ratio criteria. *Biometrika*, **43**, 295–303.

Lawless, J.F. (1982) *Statistical Models and Methods for Lifetime Data*, J. Wiley & Sons, New York.

Lawson, C.L. and Hanson, R.J. (1974) *Solving Least Squares Problems*, Prentice-Hall, Englewood Cliffs, NJ.

Lee, E.T. (1980) *Statistical Methods for Survival Data Analysis*, Lifetime Learning Publications, Belmont, CA.

Lehmann, E.L. (1959) *Testing Statistical Hypotheses*, J. Wiley & Sons, New York.

Mallows, C.L. (1973) Some comments on C_p. *Technometrics*, **15**, 661–75.

Mantel, N. and Haenszel, W. (1959) Statistical aspects of the analysis of data from retrospective studies of disease. *J. Nat. Cancer Inst.*, **22**, 719–48.

Mantel, N. and Hankey, W. (1975) The odds ratio of a 2×2 contingency table. *Am. Statistician*, **29**, 143–5.

Manton, K.G., Woodbury, M.A. and Stallard, E. (1981) A variance components approach to categorical data models with heterogeneous cell populations: analysis of spatial gradients in lung cancer mortality rates in North Carolina counties. *Biometrics*, **37**, 259–69.

Maxwell, A.E. (1961) *Analysing Qualitative Data*, Methuen, London.

McCullagh, P. (1977) A logistic model for paired comparisons with ordered categorical data. *Biometrika*, **64**, 449–53.

McCullagh, P. (1980) Regression models for ordinal data (with discussion). *J. R. Statist. Soc.*, B, **42**, 109–42.

McCullagh, P. (1982) Some applications of quasisymmetry. *Biometrika*, **69**, 303–8.

McCullagh, P. (1983) Quasi-likelihood functions. *Ann. Statist.*, **11**, 59–67.

McNemar, Q. (1947) Note on the sampling error of the difference between two correlated proportions or percentages. *Psychometrika*, **12**, 153–7.

Miller, R.G., Jr (1981) *Survival Analysis*, J. Wiley & Sons, New York.

Mosteller, F. and Tukey, J.W. (1977) *Data Analysis and Regression*, Addison-Welsey, Reading, MA.

Nelder, J.A. (1965a) The analysis of randomized experiments with orthogonal block structure. I. Block structure and the null analysis of variance. *Proc. R. Soc.*, A, **283**, 147–62.

Nelder, J.A. (1965b) The analysis of randomized experiments with orthogonal block structure. II. Treatment structure and the general analysis of variance. *Proc. R. Soc.*, A, **283**, 163–78.

Nelder, J.A. (1966) Inverse polynomials, a useful group of multi-factor response functions. *Biometrics*, **22**, 128–41.

Nelder, J.A. (1974) Log linear models for contingency tables: a generalization of classical least squares. *Appl. Statist.*, **23**, 323–9.

Nelder, J.A. (1977) A reformulation of linear models. *J. R. Statist. Soc.*, A, **140**, 48–77.

Nelder, J.A. (1982) Linear models and non-orthogonal data. *Utilitas Mathematica*, **21**B, 141–51.

Nelder, J.A. and Pregibon, D. (1983) Quasilikelihood models and data analysis. *Bell Laboratories Technical Memorandum*.

Nelder, J.A. and Wedderburn, R.W.M. (1972) Generalized linear models. *J. R. Statist. Soc.*, A, **135**, 370–84.

Oakes, D. (1977) The asymptotic information in censored survival data. *Biometrika*, **64**, 441–8.

Oakes, D. (1981) Survival times: aspects of partial likelihood. *Int. Statist. Rev.*, **49**, 235–64.

Palmgren, J. (1981) The Fisher information matrix for log–linear models arguing conditionally in the observed explanatory variables. *Biometrika*, **68**, 563–6.

Pearson, K. (1900) On a criterion that a given system of deviations from the probable in the case of a correlated system of variables is such that it can be reasonably supposed to have arisen from random sampling. *Phil. Mag.* (*5*), **50**, 157–75. Reprinted in *Karl Pearson's Early Statistical Papers* (ed. E.S.Pearson), Cambridge University Press, Cambridge (1948).

Pearson, K. (1901) Mathematical contributions to the theory of evolution. *Phil. Trans. R. Soc.*, **195**, 79–150.

Pearson, K. (1913) Note on the surface of constant association. *Biometrika*, **9**, 534–7.

Pearson, K. and Heron, D. (1913) On theories of association. *Biometrika*, **9**, 159–315.

Peto, R. (1972) Contribution to the discussion of Cox, (1972): Regression modes and life tables. *J. R. Statist. Soc.*, B, 205–7.

Pike, H.C., Casagrande, J. and Smith, P.G. (1975) Statistical analysis of individually matched case-control studies in epidemiology; factor under study a discrete variable taking multiple values. *Br. J. Prev. Soc. Med.*, **29**, 196–201.

Plackett, R.L. (1960) *Regression Analysis*, Clarendon Press, Oxford.

Plackett, R.L. (1981) *The Analysis of Categorical Data*, Griffin, London.

Pregibon, D. (1979) Data analytic methods for generalized linear models. *PhD Thesis*, University of Toronto.

Pregibon, D. (1980) Goodness of link tests for generalized linear models. *Appl. Statist.*, **29**, 15–24.

Pregibon, D. (1981) Logistic regression diagnostics. *Ann. Statist.* **9**, 705–24.

Pregibon, D. (1982) Score tests in GLIM with applications, in *Lecture Notes in Statistics*, no. 14, *GLIM.82: Proceedings of the International Conference on Generalised Linear Models* (ed. R. Gilchrist), Springer-Verlag, New York.

Pringle, R.M. and Rayner, A.A. (1971) *Generalized Inverse Matrices with Application to Statistics*, Griffin, London.

Rao, C.R. (1973) *Linear Statistical Inference and Its Applications*, 2nd edn, J. Wiley & Sons, New York.

Schoener, T.W. (1970) Nonsynchronous spatial overlap of lizards in patchy habitats. *Ecology*, **51**, 408–18.

Searle, S.R. (1971) *Linear Models*, J. Wiley & Sons, New York.

Searle, S.R. (1979) Annotated computer output for analysis of variance of unequal-subclass-numbers data. *Am. Statistician*, **33**, 222–3.

Seber, G.A.F. (1977) *Linear Regression Analysis*, J. Wiley & Sons, New York.

Simon, G. (1974) Alternative analyses for the singly ordered contingency table. *J. Am. Statist. Assoc.* **69**, 971–6.

Simpson, E.H. (1951) The interpretation of interaction in contingency tables. *J. R. Statist. Soc.*, B, **13**, 238–41.

Snell, E.J. (1964) A scaling procedure for ordered categorical data. *Biometrics*, **20**, 592–607.

Sprent, P. (1969) *Models in Regression and Related Topics*, Methuen, London.

Stevens, S.S. (1951) Mathematics, measurement and psychophysics, in *Handbook of Experimental Psychology* (ed. S.S. Stevens), J. Wiley & Sons, New York.

Stevens, S.S. (1958) Problems and methods of psychophysics. *Psychol. Bull.*, **55**, 177–96.

Stevens, S.S. (1968) Measurement, statistics and the schematic view. *Science*, **161**, 849–56.

Stewart, G.W. (1973) *Introduction to Matrix Computations*, Academic Press, New York.

Stigler, S. (1981) Gauss and the invention of least squares. *Ann. Statist.*, **9**, 465–74.

Stone, M. (1977) An asymptotic equivalence of choice of model by cross-validation and Akaike's criterion. *J. R. Statist. Soc.*, B, **39**, 44–7.

Stuart, A. (1953) The estimation and comparison of strengths of association in contingency tables. *Biometrika*, **40**, 105–10.

Stuart, A. (1955) A test for homogeneity of the marginal distributions in a two-way classification. *Biometrika*, **42**, 412–16.

Thompson, R. and Baker, R.J. (1981) Composite link functions in generalised linear models. *Appl. Statist.*, **30**, 125–31.

Tukey, J.W. (1949) One degree of freedom for non-additivity. *Biometrics*, **5**, 232–42.

Tukey, J.W. (1962) The future of data analysis. *Ann. Math. Statist.*, **33**, 1–67.

Upton, C.J.G. (1978) *The Analysis of Cross-Tabulated Data*, J. Wiley & Sons, New York.

Urquhart, N.S. and Weeks, D.L. (1978) Linear models in messy data: some problems and alternatives. *Biometrics*, **34**, 696–705.

Wampler, J.H. (1979) Solution to weighted least squares problems by modified Gram–Schmidt with iterative refinement. *ACM Trans. Math. Software*, **5**, 494–9.

Wedderburn, R.W.M. (1974) Quasi-likelihood functions, generalized linear models and the Gauss–Newton method. *Biometrika*, **61**, 439–47.

Wedderburn, R.W.M. (1976) On the existence and uniqueness of the maximum likelihood estimates for certain generalized linear models. *Biometrika*, **63**, 27–32.

Welch, L.F., Adams, W.E. and Corman, J.L. (1963) Yield response surfaces, isoquants and economic fertilizer optima for coastal Bermudagrass. *Agron. J.*, **55**, 63–7.

Whitehead, J. (1980) Fitting Cox's regression model to survival data using GLIM. *Appl. Statist.*, **29**, 268–75.

Wilkinson, G.N. and Rogers, C.E. (1973) Symbolic description of factorial models for analysis of variance. *Appl. Statist.*, **22**, 392–9.

Williams, D.A. (1976) Improved likelihood ratio tests for complete contingency tables. *Biometrika*, **63**, 33–7.

Williams, D.A. (1982) Extra-binomial variation in logistic linear models. *Appl. Statist.*, **31**, 144–8.

Williams, E.J. (1959) *Regression Analysis*, J. Wiley & Sons, New York.

Williams, O.D. and Grizzle, J.E. (1972) Analysis of contingency tables having ordered response categories. *J. Am. Statist. Assoc.*, **67**, 55–63.

Wilson, E.B. and Hilferty, M.M. (1931). The distribution of chi-square. *Proc. Nat. Acad. Sci.*, **17**, 684–8.

Yates, F. (1948) The analysis of contingency tables with groupings based on quantitative characters. *Biometrika*, **35**, 176–81.

Yule, G.U. (1912) On methods of measuring association between two attributes (with discussion). *J. R. Statist. Soc.*, **75**, 579–652.

Author index

Adams, W. E., 249
Adena, M. A., 100, 241
Agresti, A., 124, 241
Aitkin, M., 13, 183, 185, 186, 192, 241
Akaike, H., 70, 71, 241
Andersen, E. B., 100, 241
Andrews, D. F., 221, 241
Anscombe, F. J., 29, 194, 210, 241
Anturane Reinfarction Trial Research Group, 98, 241
Armitage, P., 78, 80, 100, 180, 241
Ashford, J. R., 112, 125, 241
Ashton, W. D., 100, 241
Atkinson, A. C., 69, 70, 71. 156, 220, 221, 241, 242

Baker, R. J., 199, 224, 242, 249
Barker, N. W., 246
Barndorff-Nielsen, O., 14, 29, 178, 213, 223, 242
Bartlett, M. S., 180, 238, 242
Baxter, L. A., 158, 242
Beale, E. M. L., 71, 242
Beaton, A. E., 63, 242
Belsley, D. A., 221, 242
Berge, C., 145, 242
Berk, R. H., 14, 242
Berkson, J., 11, 100, 242
Bhapkar, V. P., 124, 242
Birch, M. W., 194, 242
Bishop, Y. M. M., 25, 61, 81, 92, 100, 120, 124, 143, 147, 242
Björck, A., 67, 242
Bliss, C. I., 10, 14, 242
Bock, R. D., 125, 243
Bortkewitsch, L., 2, 243
Box, G. E. P., 6, 15, 196, 198, 243
Breslow, N., 14, 78, 100, 126, 178, 243
Bross, I. D. J., 123, 243
Byar, D. P., 89, 192, 240, 243, 244

Casagrande, J., 248
Caussinus, H., 124, 243
Chambers, E. A., 65, 76, 243
Chambers, J. M., 71, 243
Clark, V. A., 192, 246
Clarke, M. R. B., 64, 67, 243
Clayton, D. G., 13, 125, 183, 185, 186, 192, 241, 243
Coe, R., 164, 243
Cook, R. D., 220, 221, 243
Cordeiro, G. M., 223, 238, 243
Corman, J. L., 249
Coutts, S. M., 242
Cox, D. R., 14, 29, 36, 74, 75, 76, 84, 99, 100, 104, 119, 121, 126, 152, 178, 183, 188, 189, 192, 196, 198, 199, 211, 213, 221, 223, 242, 243, 244
Crilley, J., 136, 137
Crowley, J., 192, 244

Dahm, P. A., 246
Daniel, C., 96, 244
Darby, S. C., 201, 206, 207, 208, 244
Darroch, J. N., 107, 145, 148, 244
Day, N. E., 78, 89, 100, 178, 243, 244
Deming, W. E., 107, 244
Dempster, A. P., 14, 176, 224, 244
Draper, N. R., 35, 71, 244
Durbin, J., 223, 244
Dyke, G. V., 11, 244

Efron, B., 188, 244
Efroymson, M. A., 70, 71, 244
Elandt-Johnson, R. C., 192, 245
Ellis, M. J., 201, 206, 207, 208, 244
Everitt, B. S., 100, 245

Fienberg, S. E., 92, 100, 104, 107, 124, 125, 242, 245
Finney, D. J., 14, 80, 100, 125, 180, 245

Firth, D., 176, 245
Fisher, R. A., 5, 8, 9, 10, 14, 99, 100, 133, 134, 135, 136, 180, 199, 245
Fleiss, J. L., 99, 100, 245
Francis, I., 240, 245
Freireich, E. J., 186, 245
Furnival, G. M., 70, 71, 245

Gail, M. H., 90, 245
Gart, J. J., 75, 245
Gauss, C. F., 1, 7, 8, 14
Gehan, E. A., 186, 245
Gentleman, W. M., 71, 245
Givens, W. J., 66
Gokhale, D. V., 100, 148, 245
Goodman, L. A., 104, 124, 147, 242, 245
Grizzle, J. E., 125, 250
Gross, A. J., 192, 246
Gurland, J., 125, 246

Haberman, S. J., 14, 82, 100, 104, 124, 125, 144, 147, 227, 246
Haenszel, W., 89, 90, 100, 247
Hankey, W., 90, 247
Hanson, R. J., 71, 247
Harter, H. L., 71, 246
Hawkins, D., 221, 246
Healy, M. J. R., 65, 80, 180, 246
Heminway, L. N., 136, 137
Heron, D., 125, 248
Hewlett, P. S., 125, 204, 246
Hilferty, M. M., 29, 153, 250
Hinkley, D. V., 36, 84, 99, 244
Hoagslin, D. C., 220, 246
Holland, P. W., 242
Hougaard, P., 153, 246
Householder, A. S., 66, 67
Hu, M., 192, 244
Hurn, M. W., 162, 246

Jacobi, C. G. J., 67
Jenkins, G. M., 15, 243
Johnson, N. H., 192, 245

Kalbfleisch, J. D., 192, 246
Killion, R. A., 147, 246
Kruskal, W. H., 124, 246
Kuh, E., 242
Kullback, S., 100, 148, 245

Laird, N, M., 244
Lane, P. W., 18, 246
Lauritzen, S. L., 244
Lawless, J. F., 192, 247
Lawley, D. N., 86, 223, 230, 232, 237, 247
Lawson, C. L., 71, 247
Lee, E. T., 192, 247
Lee, I., 246
Legendre, A. M., 1, 7, 14
Lehmann, E. L., 126, 247
Lubin, J. H., 245

Magath, T. D., 246
Mallows, C. L., 70, 71, 247
Mantel, N., 89, 90, 100, 247
Manton, K. G., 195, 247
Maxwell, A. E., 100, 247
McCullagh, P., 104, 120, 124, 125, 171, 180, 247
McKinley, K. S., 204
McNemar, Q., 100, 247
Miller, R. G. Jr, 192, 247
Morse, P. M., 204
Mosteller, F., 71, 247

Nelder, J. A., 12, 14, 18, 26, 49, 62, 71, 154, 157, 199, 203, 212, 242, 246, 247, 248
Newell, G., 109, 110
Neyman, J., 126

Oakes, D., 183, 188, 248

Palmgren, J., 142, 248
Patterson, H. D., 11, 244
Pearson, E. S., 126
Pearson, K., 102, 109, 125, 131, 210, 248
Peto, R., 188, 190, 191, 192, 248
Pike, H. C., 124, 248
Plackett, R. L., 71, 74, 100, 120, 124, 125, 132, 147, 171, 246, 248
Poisson, S. D., 2
Pregibon, D., 89, 195, 196, 198, 212, 215, 217, 221, 223, 241, 247, 248
Prentice, R. L., 192, 246
Pringle, R. M., 63, 248

Rao, C. R., 214, 248
Ratcliff, D., 107, 244
Rayner, A. A., 63, 248
Rogers, C. E., 41, 71, 249
Ross, G. A. F., 242
Rubin, D. B., 244
Rubinstein, L. V., 245

Schoener, T. W., 92, 248

Searle, S. R., 71, 239, 248
Seber, G. A. F., 35, 71, 248
Simon, G., 125, 249
Simpson, E. H., 90, 249
Smith, H., 35, 71, 244
Smith, P. G., 248
Snell, E. J., 29, 74, 125, 211, 221, 223, 244, 249
Speed, T. P., 244
Sprent, P., 71, 249
Spurr, D. T., 204
Stallard, E., 247
Stephan, F. F., 107, 244
Stern, R. D., 164, 243
Stevens, S. S., 102, 249
Stewart, G. W., 71, 249
Stigler, S., 1, 249
Stone, M., 71, 249
Stuart, A., 120, 124, 249

Thompson, R., 224, 249
Tidwell, P. W., 199, 243
Tukey, J. W., 7, 71, 221, 247, 249

Upton, C. J. G., 147, 249
Urquhart, N. S., 62, 249

Venn, J., 46

Wampler, J. H., 71, 249
Wedderburn, R. W. M., 8, 13, 26, 29, 82, 154, 173, 175, 180, 199, 214, 248, 249
Weeks, D. L., 62, 249
Weisberg, S., 221, 243
Welch, L. F., 201, 202, 249
Welsch, R. E., 220, 242, 246
Whitehead, J., 13, 188, 191, 192, 249
Wilkinson, G. N., 41, 71, 249
Williams, D. A., 80, 180, 223, 250
Williams, E. J., 71, 250
Williams, O. D., 125, 250
Wilson, E. B., 29, 153, 250
Wilson, R. W., 70, 71, 245
Wilson, S. R., 100, 241
Woodbury, M. A., 247

Yates, F., 8, 94, 104, 250
Yule, G. U., 125, 250

Zahn, D. A., 147, 246
Zweifel, J. R., 75, 245

Subject index

Absolute error, 195
Accidentally empty cell, 61, 62, 158
Accuracy of prediction, 18
Additive effect, 23, 138, 173, 199
Additive model, 62, 149
Additivity, 16
Adjusted dependent variable, 31, 68, 70, 80, 81, 82, 172, 210, 215, 217
Algorithms
  EM, 14, 176, 177, 224, 225
  for fitting GLM, 31
  Givens, 67
  Householder, 66
  iterative proportional scaling, 81, 107, 124, 143, 239
  for least squares, 63
  maximum-likelihood, 239
  Newton–Raphson, 33, 81, 89, 172
  QR, 68
  Yates's, 94
Aliasing, 45, 46, 47, 49, 51, 52, 63, 68
  extrinsic, 51, 52
  intrinsic, 40, 43, 46, 47, 48, 52
  pseudo, 68
Analysis
  of counts, 12, 16
  of covariance, 18
  of designed experiments, 15
  of deviance, 26, 27, 28, 166, 167
  of discrete data, 100
  of dispersion, 227
  of survival data, 13, 14, 192
  of variance, 9, 26, 27, 38, 58, 59, 128, 208
Ancillary statistic, 94, 98
Anomalous value, 28
Associated counts, 60, 61, 62
Asymptotic covariance matrix, 80, 235
Asymptotic theory, 81, 82, 84, 133
Autoregressive model, 15, 222

Backward elimination, 70
Bartlett's correction factor, 238
Bayes's theorem, 78
Best subset of covariates, 70
Between-block dispersion component, 226
Between-block variance, 225
Bias, 83, 94, 112
Bilinear relationship, 38, 53
Binary data, 72, 74, 77, 100, 109, 124, 126, 166
Binomial data, 83, 86, 115, 165, 175, 177, 198, 232
Binomial denominator, 73, 74, 79, 82, 83
Binomial distribution, 2, 13, 22, 23, 24, 25, 26, 29, 69, 73, 74, 85, 119, 124, 128, 129, 140, 141, 154, 170, 171, 194, 204, 212, 214, 216, 232
Binomial model, 124, 143, 176
Bioassay, 10, 11, 133, 134, 204, 207, 208
Bivariate binary response, 145
Block factor, 183, 191
Block parameter, 225

Calibration, 18
Canonical form of log-likelihood, 32
Canonical scale, 20, 21, 24, 25, 33, 88, 154, 191, 194, 195, 202
Cardinal scale, 102
Causal relationship, 144
Censoring, 13, 181, 183, 184, 186, 187, 189, 239
Central-limit theorem, 36, 236
Checking
  a covariate scale, 211, 217
  the link function, 212, 218
  the variance function, 212, 218
Chi-squared approximation for deviance, 28, 111

Chi-squared distribution, 26, 29, 86, 156, 171, 229, 230, 231, 234
Choleski decomposition, 63, 65, 66
Classical linear model, 1, 2, 4, 7, 8, 11, 17, 18, 19, 20, 23, 27, 35, 36, 54, 63, 149, 193, 199, 239
Classifying factor, 38, 61, 62, 136, 160
Closed-form estimate of parameter, 145
Coefficient of variation, 13, 149, 156, 157, 158, 160, 166, 203
Combinatorial explosion, 69, 221
Combined estimate, 119
Combining response categories, 103, 112
Comparison
 of Poisson means, 141
 of variance functions, 214
Components of variance, 151, 156, 157, 180, 225, 227
Composite hypothesis, 235, 238
Compound term, 40, 41, 42, 43, 44
Computer packages, 40, 41, 143, 239
Condition number, 65
Conditional covariance matrix, 89
Conditional distribution, 87, 89, 117, 121, 122, 123, 124, 141, 147
Conditional expectation, 118, 119, 224, 226
Conditional fit, 12
Conditional independence, 144
Conditional maximum-likelihood estimate, 89, 90, 91, 120
Conditional mean, 88, 90, 178
Conditional probability, 99, 104, 105, 123, 146, 147, 189, 190
Conditional variance, 90, 117, 226
Confidence limit, 84, 85, 116, 133, 135, 139, 173, 179, 187, 197, 201, 206
Confounded effect, 116
Consistency, 90, 91, 109, 118, 133, 142, 149, 180, 226, 234
Constancy of variance, 7, 19, 199, 206
Constant-information scale, 216
Constraints, 48, 49, 51
 conventionality of, 49, 51
 symmetric, 50
Constructed variable, 209, 221
Contingency table, 104
Continuation-ratio model, 104, 120, 125
Continuity correction, 74, 92, 99
Continuous covariate, 38, 39, 40, 41, 42, 54
Continuous scale, 125, 149, 212
Contrast, 65, 94, 97, 136, 145
Convergence
 of iteration, 68, 82, 123, 193, 196, 200, 215
 to Normality, 74
Convolution, 151
Cook's statistic, 220, 221
Correlation, 16, 111, 112, 162, 176, 204, 206, 236
 matrix, 110, 117, 203
Counts, 2, 9, 11, 13, 23, 28, 61, 112, 125, 144, 209, 239
Covariance of estimates, 64, 80, 87, 108, 109, 118, 119, 120, 133, 156, 157, 168, 171, 193, 194, 195, 200, 201, 220, 236
Coverage probability, 84
Cox's proportional-hazards model, 183, 188
Cross-classification, 11, 39, 49, 51
Crossing factor, 142
Cumulant, 74, 84, 128, 150, 229, 231, 237, 238
 fourth, 173
 third, 223
Cumulative distribution, 129, 171
Cumulative probability, 103, 105, 107, 117, 125
Cycle of graph, 145
Cyclic hypergraph, 145
Cyclical term, 165

Data matrix, 7, 60, 61, 210, 239
Decomposable model, 144, 145, 148
Design
 D-optimum, 220
 balanced, 162
 experimental, 8, 18, 39, 72, 181, 225
 factorial, 8
 Latin square, 134
 one-at-a-time, 8
 orthogonal, 156
 split-plot, 16, 225
Deviance, 17, 24, 25, 26, 27, 28, 30, 68, 69, 74, 81, 82, 86, 97, 111, 112, 115, 139, 160, 161, 162, 163, 164, 166, 167, 172, 178, 186, 187, 195, 196, 197, 202, 204, 206, 207, 211, 212, 213, 215, 236, 238
 component, 230, 231, 232
 function, 234
 test, 214, 215

Dilution assay, 9
Dimensionality of subspace, 45, 46, 47
Direct decomposition, 65, 67
Disconnectedness, 52
Discrepancy of a fit, 4, 5, 24, 25, 26, 27, 56, 69
Discrete covariate, 38
Discrete data, 125, 168, 180, 212, 216, 217, 218
Discrete distribution, 105, 213, 214
Dispersion parameter, 21, 31, 56, 63, 69, 83, 108, 127, 133, 157, 171, 172, 193, 194, 204, 212, 213, 220
Dispersion, component of, 136, 225
Distribution of the deviance, 28, 83, 229, 231, 233, 237
Divergence of iteration, 68
Dose, 18
Dummy variable, 40, 41, 42, 49, 54
Dummy vector, 45, 46, 47, 48

ECTA, 239
Educational testing model, 100
Effective degree of freedom, 86, 109, 133, 173
Effective sample size, 161
Embedding a covariate (in a family), 209
Empty cell, 61
Envelope, 220
Enzyme kinetics, 12
Epidemiological research, 100
Erlangian distribution, 152
Errors with constant variance, 13, 35
Estimable quantity, 48, 50, 51, 63
Estimator of the odds ratio, 91
Euclidean space, 45
Exact test, 89
Existence of parameter estimate, 82
Expected Normal order statistic, 216
Explanatory variable, 37, 72, 100, 105
Exploratory analysis, 144, 150
Exponential distribution, 2, 28, 150, 167, 171, 182, 185, 186
Exponential family, 14, 20, 22, 122, 169, 171, 175, 189, 190, 191, 213
Exponentially weighted sampling, 190
Extended quasi-likelihood, 212
Extensive variable, 150
Extra parameter, 209
Extreme-value distribution, 76, 186

F-distribution, 69, 201
F-ratio, 69, 139, 207
Factor, 7, 8, 9, 16, 26, 39, 40, 41, 46, 47, 48, 70, 72, 100, 162
Factor level, 39, 40, 41, 42, 45, 61, 62, 201
Factorial contrast, 145, 146, 147
Factorial experiment, 8, 9, 201
Factorial model, 8, 9, 11, 26, 43, 45
Failure time, see Survival time
False positive, 69
Faulty observation, 109
Finite population, 190
First-order Markov model, 165, 166
First-order dependence, 164
Fisher's exact test, 99, 100
Fisher's scoring method, 14, 32, 33, 125
Fitted marginal total, 54
Fitted value, 4, 5, 10, 17, 28, 31, 54, 62, 69, 70, 82, 175, 196, 203, 208, 217, 219, 238
Fixed margin, 99
Forward selection, 70
Fourier series, 166
Fourth-moment assumption, 237, 238
Frequency distribution, 17, 105
Full model, 25, 27
Functional marginality, 53, 70
Functional relations in covariates, 52
Fused cells, 62, 224

G^2 statistic, 25
Gamma deviance, 25, 26
Gamma distribution, 2, 22, 24, 25, 29, 37, 132, 150, 151, 152, 153, 155, 156, 158, 162, 163, 164, 170, 171, 194, 195, 202, 203, 212, 216, 217
Gamma index, 150
Gaussian distribution, see Normal distribution
Gaussian elimination, 63
GENCAT, 240
Generalized extreme-value distribution, 186
Generalized inverse matrix, 63, 65
Genstat, 42, 45, 239
Geometrical interpretation of least squares, 54
Geometry of the score test, 215
GLIM, 42, 45, 51, 239
Goodness-of-fit statistic, 17, 24, 26, 28, 55, 60, 85, 105, 115, 160, 161, 195
Goodness-of-link test, 195, 221
Grand mean, 48, 63, 227
Graph theory, 144, 145, 148
Graphical analysis, 105, 192, 214

Grouped data, 73, 85, 109, 190, 224

Half-normal plot, 96, 220
Harmonic term, 165, 166, 167
Hazard function, 181, 182, 183, 184, 186, 188, 189
Hazard ratio, 187
Health sciences, 100
Hereditary hypergraph, 145
Hessian matrix, 32, 33, 34
Hierarchical model, 145
Histogram, 164
Hyperbolic response, 12, 162, 201, 204
Hypergeometric distribution, 99, 117, 121, 129, 177

Incidence matrix, 156
Independence, 4, 19, 79, 144
Independent observation, 15, 17, 73, 115
Indicator variable, 72
Influential point, 209, 210, 218, 220, 221
Information, 8, 10, 46, 55, 64, 69, 143
  combination of, 91
  Fisher's, 5, 34, 56, 234
  loss of, 117
  matrix, 60, 63, 68, 108, 132, 133, 135, 141, 143, 171, 187, 233
Inner-product matrix, 169
Instantaneous risk, 182
Interaction, 26, 39, 43, 51, 95, 96, 98, 128, 135, 138, 139, 140, 144, 146, 147
  first-order, 39
  three-factor, 96
  two factor, 9, 39, 97, 138, 161, 203
Intercept, 10, 38, 40, 52, 63, 78, 149, 161, 162, 184, 204, 206
Interval scale, 102
Inverse Gaussian distribution, 22, 24, 25, 28, 29, 170, 171, 212, 213
Inverse linear response, 154, 155, 162, 201, 203
Inverse polynomial, 12, 154, 155, 201, 203
Inverse scale, 212, 216
Iterative weighted least squares, 14, 31, 68, 81, 107, 133, 150, 180

Kurtosis, 84, 109

Least squares, 1, 8, 55, 57, 58, 60, 66, 80, 149, 169
Likelihood
  binomial, 79, 80
  conditional, 87, 88, 100, 116, 123, 124
  function, 5, 17, 20, 25, 29, 36, 53, 73, 79, 81, 84, 88, 106, 107, 108, 116, 128, 130, 141, 152, 153, 165, 171, 175, 176, 177, 178, 183, 184, 185, 186, 190, 191, 194, 201, 212, 215, 237, 238
  gamma, 152, 157
  hypergeometric, 165, 177, 178
  multinomial, 106, 107, 109, 141, 142, 191, 192
  negative binomial, 132
  Poisson, 130, 131, 132, 138, 141, 142, 191, 192, 232
Likelihood-ratio statistic, 82, 84, 87, 92, 109, 112, 115, 152, 213, 215, 223, 237, 238
Linear dependency, 51
Linear model, 15, 46, 75, 88, 103, 122, 149, 156, 160, 215
Linear predictor, 11, 19, 20, 23, 28, 31, 35, 38, 40, 41, 42, 44, 45, 46, 48, 49, 50, 51, 52, 54, 68, 69, 163, 183, 184, 197, 199, 204, 210, 211, 212, 213, 217, 239
Linear regression, 91, 99, 128, 211, 224
Linear trend, 180
Linearization, 199, 201, 204
Linearly independent vectors, 46
Link
  complementary log–log, 23, 76, 77, 103, 175, 198, 204
  composite, 14, 223, 224
  function, 19, 20, 23, 31, 35, 76, 77, 103, 155, 156, 176, 184, 193, 195, 196, 197, 198, 199, 209, 211, 212, 216, 233, 239
  identity, 20, 23, 35, 53, 155, 198, 212
  log, 23, 149, 155, 162, 197
  logit, 11, 13, 23, 76, 77, 80, 96, 115, 146, 198, 204
  non-canonical, 24, 68
  parameter, 196, 198
  probit, 23, 76, 77, 103, 115, 125, 204
  reciprocal, 160, 162, 197, 202, 213, 217, 219
Log dose, 204
Log odds, 11, 103
  ratio, 99, 178
Log scale, 135, 149, 162
Log–linear model, 11, 12, 13, 106, 107, 121, 122, 124, 125, 126, 127, 128, 134, 136, 140, 142, 143, 144, 145, 146, 147, 148, 184, 185, 186, 191, 239

Log-normal distribution, 150
Logistic model, 11, 78, 96, 97, 100, 143, 147, 165, 198
Logistic regression, 225
Logistic scale, 78, 81, 94, 146
Loss of dimension, 45
Lower-triangular matrix, 108

Main-effect, 9, 39, 95, 128, 138, 143, 146, 160, 161, 197, 212, 216, 217, 219
Main-effects model, 138, 139
Mantel–Haenszel estimator, 91
Mantel–Haenszel test, 89
Marginal distribution, 125, 147
Marginal frequency, 175, 176
Marginal homogeneity, 124
Marginal relation, 48, 50, 51, 52
Marginal total, 16, 54, 89, 90, 92, 112, 117, 122, 129, 140, 178
Marginality, 70, 125
Markov chain, 164
Matched pairs, 90, 100, 119, 123, 124, 125
Matched triplets, 123, 124
Maximum likelihood, 10, 25, 96, 148, 195, 225, 239
  equation, 32, 33, 53, 54, 123, 164, 194, 195
  estimate, 5, 10, 14, 26, 31, 53, 54, 55, 56, 42, 144, 154, 157, 158, 177, 196, 213, 224, 225, 234
Mean-value parameter, 17, 25, 89, 168
Measure of association, 124
Measurement error, 1
Measurement scale, 101, 102, 147
Michaelis–Menten equation, 12
Minimum-discrimination information, 148
Missing value, 224
Mixed term, 40, 41
Mixture of drugs, 200, 206
Model
  for binary responses, 75
  checking, 6, 87, 209, 210, 214, 221
  for dependence, 145, 147
  fitting, 4, 5, 6, 13, 41, 56
  formula, 41, 42, 43, 44, 45, 48, 144, 147, 239
  identification, 15
  for nominal scales, 105
  for ordinal scales, 102
  for survival data, 181
Moment, 74, 75, 86, 122, 150, 177, 231
Multi-way table, 16, 60
Multinomial distribution, 12, 13, 107, 108, 128, 140, 141, 171, 227
Multinomial index, 132, 151, 153, 156
Multinomial observation, 106, 107, 108, 111, 117, 120, 121, 145, 146, 191
Multinomial response, 106, 116, 128, 140, 142, 143, 144, 145, 146
Multiple responses, 144
Multiplicative model, 11, 16, 23, 88, 134, 138, 149, 156, 162, 166
Multivariate Normal density, 225
Multivariate hypergeometric distribution, 122

Necessarily empty cell, 61, 137, 138
Negative binomial distribution, 132, 170, 171, 194, 195, 212, 214
Nested sets of models, 26, 171
Nested structure, 225
Nesting index system, 43
Nominal response, 102, 106
Non-central hypergeometric distribution, 87, 171, 189
Non-linear parameter, 105, 193, 199, 200, 202, 204, 206
Non-linearity, 38, 113, 193, 199
Non-orthogonality, 27
Normal distribution, 1, 2, 4, 7, 11, 19, 20, 22, 24, 25, 26, 28, 36, 37, 53, 74, 170, 171, 193, 211, 212, 213, 225, 235
Normal equation, 63
Normal integral, 10, 76
Normal limit, 129, 151
Normal model, 28, 70, 169, 210, 216, 221
Normal variable, 19, 156, 220
Normality, 16, 36, 199, 203
Nuisance parameter, 16, 78, 87, 100, 116, 117, 119, 121, 122, 124, 125, 143, 223, 235, 238
Null distribution, 85, 92
Null hypothesis, 125, 141
Null model, 25
Null vector, 42

Observational data, 8, 9, 72, 181
Odds, 106, 114, 115
Odds ratio, 76, 87, 88, 89, 91, 98, 121, 122, 165, 179
Offset, 138, 184, 185, 187
Operators
  +, 42
  crossing, 43, 44
  dot, 42

Operators (*cont.*)
exponential, 45
for removal of terms, 44
nesting, 43, 44
Optimum model, 17
Ordering of terms, 48, 50, 53
Ordinal measurement scale, 102, 117
Ordinal regression model, 112
Ordinal response, 102, 108
Orthogonal data, 26
Orthogonal matrix, 65, 67
Orthogonal parameters, 58
Orthogonality, 48, 59, 67
Outer product, 40
Outlier, 6, 208, 218
Over-dispersion, 69, 73, 80, 83, 84, 86, 94, 96, 106, 108, 127, 131, 132, 138, 139, 180, 232

Parameter in the link function, 195
Parameter in the variance function, 193, 195
Parameter invariance, 6
Parameter space, 171, 176
Parametric survival functions, 188, 192
Parsimony, 6, 27, 68, 69, 128
Partial differential equation, 168, 169
Partial likelihood, 121, 126, 188, 189, 191, 192
Path model, 144
Pattern in data, 2
Pattern in residuals, 210
Pearson chi-squared statistic, 25, 29, 83, 85, 86, 109, 172, 194, 215, 226
Pivot, 64, 65
Poisson approximation, 74
Poisson deviance, 25, 26, 130
Poisson distribution, 2, 9, 11, 12, 13, 16, 22, 23, 24, 25, 26, 28, 29, 30, 69, 74, 128, 129, 132, 138, 154, 170, 171, 194, 212, 214, 216
Poisson model, 127, 131, 139
Poisson observation, 140, 141, 177, 184, 230, 232
Poisson process, 127, 131, 132, 152, 182
Poisson series, 9
Poisson variance function, 128, 194
Polychoric correlation coefficient, 125
Polynomial, 12, 38, 52, 70, 88, 89, 90, 128, 155, 177, 178
Polytomous data, 101, 107, 124
Positive-definitive matrix, 63, 108, 168
Potency, relative, 135, 136
Power family, 23, 196, 198
Prediction, 6, 15, 18, 69, 70
Principal axis, 60
Prior value, 138, 211
PRISM, 239
Probability distribution, 4, 20, 115, 128, 189
Probit analysis, 10, 13, 14, 198
Projection matrix, 55, 57, 210, 220
Proportion, 2, 11, 13, 80, 128, 140, 173
Proportional error, 195
Proportional-hazards model, 120, 183, 189
Proportional-odds model, 117, 122, 125, 126
Prospective study, 77, 78
Pseudo-Poisson model, 192

Quadratic polynomial, 53, 203
Quadratic variance function, 149, 227
Quadratic weight function, 31, 80, 156, 191
Qualitative covariate, 7, 38, 39, 40, 70, 138, 145,
Quantal-response assay, 18
Quantitative covariate, 7, 72, 88, 112, 138, 145
Quasi-likelihood, 8, 13, 14, 73, 80, 81, 106, 107, 108, 109, 168, 169, 170, 171, 175, 176, 177, 180, 194, 212, 213, 222, 233, 236, 237, 238
Quasi-likelihood equation, 169, 177, 234
Quasi-likelihood estimate, 176, 177, 213, 238
Quasi-likelihood model, 168, 213, 214
Quasi-likelihood-ratio statistic, 171
Quasi-symmetry, 124

Random component, 13, 19, 25, 35, 150, 199
Random sample, 141
Random variable, 117, 121, 124, 177
Random variation, 101, 138
Rank, 122
Rating scale, 101
Ratio scale, 102
Recording error, 210
Reference set, 140
Reflection, 66
Regression
analysis, 15, 16, 60, 64, 67
coefficient, 138, 168, 217, 235
diagnostic, 210

Regression (*cont.*)
model, 9, 11, 38, 209, 210, 211
with empirical weights, 239
Residual, 6, 7, 28, 36, 67, 85, 86, 87, 92, 109, 111, 113, 115, 130, 138, 161, 169, 179, 180, 203, 204, 206, 208, 210, 218, 219, 220
Anscombe, 28, 29, 30, 210, 216
cross-validatory, 220
cumulative, 113, 114
deviance, 28, 30, 86, 210, 211, 216, 217, 223
jack-knife, 220
lower bound for, 216
partial, 217
Pearson, 28, 29, 30, 31, 86, 87, 163, 210, 216
plot, 112, 216, 218
sum of squares, 25, 26, 27, 58, 59, 154, 172, 201, 206
standardized, 97, 114, 138, 139, 161, 175, 210, 211, 216, 220, 221
variance, 56, 69, 210
vector, 55, 59
Response, 69, 100
category, 103, 104, 112, 113, 118, 122, 125
factor, 12
rate, 18
surface, 201
variable, 7, 94, 101, 147
Retrospective study, 77, 91, 100, 125, 179
Ridit score, 123
Risk, 114, 115, 116, 138
Risk set, 115, 121, 188, 189, 190, 191
Robustness, 222
Rotation, 66
Rounding error, 65

Saddle point, 53, 213
Sample space, 54
Sampling error, 56, 200
Scale of measurement, 16, 103
Scaled chi-squared, distribution, 161
Scaling problem, 16
Scope of model, 6
Score, 104
Score statistic, 84, 89, 171, 195, 214
Second-derivative matrix, 172
Second-moment assumption, 36, 79, 80, 109, 169, 222, 233, 234, 235, 236
Secular trend, 166
Selection of covariates, 15, 16, 17, 26, 27, 68, 160, 161, 162
Selection, allowance for, 96, 216
Significance level, 28, 98, 99, 141
two-sided, 92, 98, 99
Significance test, 92, 141, 223
two-sided, 223
Significant difference, 162
Similar region, 126
Simplicity of model, 6
Simpson's paradox, 91
Simulation experiment, 112
Skewness, 129, 151
Slope, 10, 38, 39, 40, 41, 204, 216
Small-sample theory, 36
Software, 239
Solution locus, 54, 55, 56, 59
Stability of estimates, 59
Standard deviation, 4, 28, 36, 37, 124, 202
Standardization, 18
Standardized contrast, 95, 96
Starting procedure, 32
Starting value, 193, 196, 200, 202, 204
Statistical model, 102
Stepwise regression, 70
Stimulus variable, 7, 37, 209
Stirling approximation, 213, 214
Stochastic ordering, 104, 105
Stochastic process, 164
Straight-line relation, 3
Stratifying factor, 178
Structural zero, 61
Subspace, 45, 46, 47, 52
Successive orthogonalization, 66, 67
Sufficient statistic, 24, 81, 89, 92, 107, 122, 140, 150, 154, 224
Sum of exponentials, 200
Sum of squares, 5, 26, 54, 56, 57, 59, 64, 151, 156
Survey data, 39, 61, 68, 195
Survival analysis, 121, 186
Survival data, 13, 100, 104, 121, 181, 239
Survival distribution, 181, 182, 183, 184, 186
Survival time, 121, 181, 182, 188, 189, 190, 191, 199
Sweep operator, 63, 64, 67
Systematic component, 3, 18, 19, 25, 35, 37, 79, 81, 82, 88, 101, 134, 149, 168, 173, 199, 200, 219, 238
Systematic effect, 2, 11, 13, 16, 24, 160

t-distribution, 69, 84, 133, 135, 173

Table 2 × 2, 87, 88, 91, 98, 165, 178
Tables as data, 60, 61, 91, 239
Tail probability, 129, 130, 141
Taylor series, 75, 122, 131, 196
Test statistic, 104, 235
Tetrachoric correlation coefficient, 125
Theory as pattern, 3
Threshold value, 204
Ties
  Cox's method, 188, 189, 190, 192
  Peto's method, 188, 190, 191
Time series, 15, 18
Time-dependent covariate, 181, 239
Tolerance distribution, 198
Total sum of squares, 58
Transcription error, 210
Transformation
  angular, 74, 173, 175
  cube-root, 29, 153
  of data, 196, 198
  of fitted values, 198
  linear, 108
  to linearity, 105
  log, 135, 175, 199
  log–log, 103, 175
  logistic, 75, 104, 110, 125, 146, 173, 175
  non-linear, 221
  normalizing, 152
  power family, 129
  probit, 110
  reciprocal, 160, 161, 199
  square-root, 16, 199, 206, 220
  to symmetry, 74
  variance-stabilizing, 74, 129, 156
Transformed variable, 113
Transition probability, 176
Treatment contrast, 135
Treatment factor, 39, 110
Tree-search methods, 70
Triangular matrix, 65
Two-way table, 49, 62, 104, 142

Unbiasedness, 8, 226
Unbounded estimate, 68
Uncensored observation, 184, 186
Unconditional estimate, 119
Unconditional mean, 90, 178
Uncorrelated observation, 15
Under-dispersion, 73, 80, 84, 127, 132, 232
Ungrouped data, 73, 79, 109
Uniqueness of parameter estimate, 82
Upper-triangular matrix, 65, 66, 67
User-defined contrast, 239

Variance, 4, 29, 134, 149, 173
  function, 2, 21, 22, 25, 26, 28, 31, 153, 169, 170, 171, 173, 175, 191, 193, 194, 197, 199, 211, 212, 213, 214, 216, 218, 219
  of parameter estimate, 108
Vector of means, 18
Vector space, 45
Venn diagram, 46
Visual display, 209

Weibull distribution, 185, 186, 187
Weibull model, 188
Weight, 21, 33, 34, 61, 68, 70, 151, 153, 156, 160, 166, 168, 215
Weighted least squares, 33, 132, 151, 168, 171, 190, 224
Weighting matrix, 82, 83, 108, 157, 172, 175
Within-block dispersion component, 226
Within-block variance, 225
Within-groups analysis, 16